ISBN 978-0-266-65291-5
PIBN 10876252

58TH CONGRESS, }   HOUSE OF REPRESENTATIVES.   } DOCUMENT
2d Session.   }   } No. 710.

Professional Paper No. 22   Series H, Forestry, 7

DEPARTMENT OF THE INTERIOR

UNITED STATES GEOLOGICAL SURVEY

CHARLES D. WALCOTT, DIRECTOR

FOREST CONDITIONS

IN THE

SAN FRANCISCO MOUNTAINS FOREST RESERVE, ARIZONA

BY

JOHN B. LEIBERG, THEODORE F. RIXON, AND ARTHUR DODWELL

WITH AN INTRODUCTION BY

F. G. PLUMMER

WASHINGTON

GOVERNMENT PRINTING OFFICE

1904

Apl. 1908
18043

CONTENTS.

ILLUSTRATIONS.

LETTER OF TRANSMITTAL.

DEPARTMENT OF THE INTERIOR,
UNITED STATES GEOLOGICAL SURVEY,
Washington, D. C., October 10, 1903.

SIR: I transmit herewith, for publication in the series of Professional Papers, a report on forest conditions in the San Francisco Mountains Forest Reserve, Arizona, prepared by Mr. F. G. Plummer from the preliminary report of Mr. John B. Leiberg and notes by Messrs. Theodore F. Rixon and Arthur Dodwell.

Very respectfully,

HENRY GANNETT,
Geographer.

Hon. CHARLES D. WALCOTT,
Director United States Geological Survey.

FOREST CONDITIONS IN THE SAN FRANCISCO MOUNTAINS FOREST RESERVE, ARIZONA.

By JOHN B. LEIBERG, THEODORE F. RIXON, and ARTHUR DODWELL.

INTRODUCTION.

By F. G. PLUMMER.

BOUNDARIES.

By proclamation of President McKinley dated August 17, 1898, the San Francisco Mountains Forest Reserves were created in the Territory of Arizona, being described by sections of land as follows:

"The even-numbered sections in townships twenty-five (25), twenty-four (24), and twenty-three (23) north, ranges three (3) to nine (9) east, both inclusive; townships twenty-two (22) and twenty-one (21) north, ranges one (1) to nine (9) east, both inclusive; townships twenty (20) and nineteen (19) north, ranges one (1) to ten (10) east, both inclusive; townships eighteen (18) and seventeen (17) north, ranges four (4) to eleven (11) east, both inclusive; township sixteen (16) north, ranges five (5) to eleven (11) east, both inclusive; sections two (2), four (4), six (6), eight (8), ten (10), twelve (12), fourteen (14), sixteen (16), and eighteen (18), township fifteen (15) north, range six (6) east; sections two (2), four (4), six (6), eight (8), ten (10), twelve (12), fourteen (14), sixteen (16), eighteen (18), twenty (20), twenty-two (22), and twenty-four (24), township fifteen (15) north, range seven (7) east; sections two (2), four (4), six (6), eight (8), ten (10), twelve (12), fourteen (14), sixteen (16), eighteen (18), twenty (20), twenty-two (22), and twenty-four (24), township fifteen (15) north, range eight (8) east; sections two (2), four (4), six (6), eight (8), ten (10), twelve (12), fourteen (14), sixteen (16), and eighteen (18), township fifteen (15) north, range nine (9) east; sections two (2), four (4), and six (6), township fifteen (15) north, range ten (10) east, and sections two (2), four (4), and six (6), township fifteen (15) north, range eleven (11) east."

On April 12, 1902, President Roosevelt issued a proclamation "for the purpose of consolidating into one reserve the lands heretofore embraced in the San Francisco Mountains Forest Reserves and of including therein the other adjacent lands within

11

the description hereinafter given." The consolidated area is now known as the San Francisco Mountains Forest Reserve, and is described by metes and bounds as follows:

"Beginning at the northwest corner of township twenty-two (22) north, range one (1) east, Gila and Salt River meridian, Arizona; thence southerly along the said meridian, allowing for the proper offset on the fifth (5th) standard parallel north, to the southwest corner of Township nineteen (19) south, range one (1) east; thence easterly along the surveyed and unsurveyed township line to the point for the northwest corner of township eighteen (18) north, range four (4) east; thence southerly along the unsurveyed range line to its intersection with the fourth (4th) standard parallel north; thence easterly along said parallel to the point for the northwest corner of township sixteen (16) north, range five (5) east; thence southerly to the point for the southwest corner of said township; thence easterly to the point for the northwest corner of township fifteen (15) north, range six (6) east; thence southerly to the point for the southwest corner of section eighteen (18), said township; thence easterly along the unsurveyed section line to the point for the northwest corner of section nineteen (19), township fifteen (15) north, range seven (7) east; thence southerly to the southwest corner of said section; thence easterly along the unsurveyed section lines to the southwest corner of section nineteen (19), township fifteen (15) north, range nine (9) east; thence northerly to the northwest corner of said section; thence easterly along the section line to the southeast corner of section thirteen (13), said township; thence northerly to the northeast corner of section twelve (12), said township; thence easterly along the section lines to the southeast corner of section one (1), township fifteen (15) north, range eleven (11) east; thence northerly along the range line to its intersection with the fourth (4th) standard parallel north; thence westerly along said parallel to the southeast corner of township seventeen (17) north, range eleven (11) east; thence northerly along the surveyed and unsurveyed range line to the point for the northeast corner of township eighteen (18) north, range eleven (11) east; thence westerly to the southeast corner of township nineteen (19) north, range ten (10) east; thence northerly along the range line to its intersection with the fifth (5th) standard parallel north; thence westerly along said parallel to the point for the southeast corner of township twenty-one (21) north, range nine (9) east; thence northerly along the unsurveyed range line, allowing for the proper offset on the sixth (6th) standard parallel north, to the point for the northeast corner of township twenty-five (25) north, range nine (9) east; thence westerly along the surveyed and unsurveyed township line to the point for the northwest corner of township twenty-five (25) north, range three (3) east; thence southerly along the surveyed and unsurveyed range line, allowing for the proper offset on the sixth (6th) standard parallel north, to the northeast corner of township twenty-two (22) north, range two (2) east; thence westerly along the township line to the northwest corner of township twenty-two (22) north, range one (1) east, to the place of beginning."

Mr. John B. Leiberg began the field work in this reserve in the early part of June, 1901, and by December 1 had examined thirty-seven townships in the northern portion and five and one-half townships in the southwestern portion.

During the winter of 1901–2 Mr. Leiberg prepared a full detailed preliminary report on these areas, which are outlined on the key diagram on Pl. I.

The examination was concluded during the season of 1902 by Mr. Theodore F. Rixon and Mr. Arthur Dodwell.

For the purpose of publishing the results of these examinations as one professional paper, Mr. Leiberg's report has been supplemented by the data collected by Messrs. Rixon and Dodwell. To avoid unnecessary repetition, and for simplicity of reference, each topic or subhead and each table contains the results of the examining parties in the same manner as if the field work had been in charge of one person, and the entire area treated by one author.

Each section of 640 acres was separately cruised and reported upon in detail. The Michigan practice was used in estimating the timber, and includes trees which have a diameter as small as 8 inches at a height of 2 feet from the ground. The practice of Arizona lumbermen will average about 70 per cent of the Michigan practice estimates as given in this report.

In the preparation of the land classification map, Pl. 1, it was not attempted to show barren tracts which naturally support neither woodland nor forest. The larger tracts aggregate 362,075 acres, but the actual area is considerably greater than these figures imply, as numerous lanes and irregular openings interrupt the woodland growth. The map is a graphic representation of the existing conditions, which are set forth in greater detail in the following report.

SURFACE FEATURES.

This reserve comprises portions of the broad summit, and in localities also the slopes of an elevated tract of land in north-central Arizona. The altitude ranges from 3,500 feet at Oak Creek in the southwestern portion, to 12,794 feet at the summit of San Francisco Peak. In a general way the region may be regarded as a plateau—an extension southward of the Colorado Plateau—gashed by numerous deep canyons in the southern part and dotted by several hundred volcanic cones, principally in the northern areas. The general slopes are toward the northeast and southwest from a sinuous and ill-defined crest line which traverses the country from northwest to southeast. The northeast slope has a low gradient and leads to the valley of Little Colorado River; the southwestern slopes are steeper and at last break off precipitously into the valley of Verde River.

The plateau originated in the uplift of a vast block of territory. Volcanic vents opened in many localities over its surface and poured out great floods of lava, after which from many craters were ejected vast masses of cinders and pumice-like scoriæ, all of which helped to build up the plateau to its present level. Most of the region is therefore buried under various forms of volcanic ejecta, and only in occasional localities and in the bottoms of the deeper canyons do the underlying limestones outcrop.

The top of the plateau consists, in a general way, of rolling land. The high portion of the region is composed mostly of terraces, which mark successive flows of lava from many points of eruption. Except where the plateau breaks off to the Verde drainage, these terraces terminate in comparatively low fronts, which are seldom more than 300 feet in height, more frequently not over 50 to 100 feet. Their slopes are often steep and abrupt, and mostly rocky and talus strewn. The different volcanic foci scattered over the plateau, which accentuate the high relief, consist of (1) low, truncated, more or less symmetrical cones, built up of loose lava, scoriæ, or of cemented volcanic tuff; (2) large craters around which in the course of many successive eruptions there accumulated lava, cinder, and slag until the masses attained the height and proportions of mountains; (3) irregular ridges composed of various sorts of volcanic matter, rarely exceeding a mile in length and 1,200 feet in height; and (4) blocks of limestone, sometimes 1,500 feet in height and 3,800 feet in diameter.

The cones vary in height from 300 to 1,000 feet, and even attain 1,200 feet in one or two instances. Their diameters, at the base, vary from 1,400 to 4,500 feet. Their summits appear flat and truncate, when viewed from a distance, but a near inspection usually reveals a central depression, sometimes in the shape of an inverted cone, sometimes a shallow dish-like concavity. On many of the cones the depression is situated slightly to one side or occurs far down on the slopes.

The centers of eruption around which sufficient lava and cinders accumulated to form mountains, are San Francisco Mountains, Kendrick Peak, Sitgreaves Peak, Bill Williams Mountain, Mormon Mountain, and Apache Maid. The last-named is a very prominent peak, although the lowest of those mentioned. San Francisco Mountains and Kendrick Peak were not built up from the ejecta of one crater, but from the outpourings of many; several craters, in the case of San Francisco Mountains, eventually coalescing and causing the formation of one large central cone, while in Kendrick Peak they remain separate, grouped on either side of a central ridge. Sitgreaves Peak, on the other hand, appears to have been built up by lava flows welling out through fissures, and the ancient craters, if any existed, are not clearly traceable.

The low portion of the region consists of runs, gullies, ravines, and canyons. In the northern areas of the reserve these depressions are shallow and unimportant, except such as head in the high mountains. In the southern portion of the reserve are three principal canyons with numerous tributaries. Sycamore and Oak Creek canyons have high precipitous walls, in places aggregating 2,000 feet of inaccessible cliffs. Beaver Creek Canyon is similar, but the slopes are not as abrupt and are generally free from rock bluffs and cliffs.

SOIL.

The principal classes of soil in the reserve are as follows: Gravelly loam, loamy gravel, scoriaceous soil, cinder, and adobe.

The gravelly loam is the prevailing type of soil. It consists of loamy or adobe constituents mixed with varying percentages of volcanic débris in the areas covered with lava, and with limestone fragments when outcrops of that rock occur. The loamy gravel occurs in the limestone belt and in the northeast corner of the reserve, where it abuts on the desert areas of the Little Colorado. The scoriaceous soils prevail on the slopes of the volcanic cones and ridges and in their immediate neighborhood. They are distributed throughout the reserve, and are composed of a mixture of slag and pumice-like detritus, red, gray, or yellow, with small quantities of loamy matter intermixed. The cinder soils occur in situations similar to those of the scoriaceous ones, but are especially typical of the volcanic areas lying between the eastern foot of San Francisco Peak and the Little Colorado Desert. This class of soil consists almost wholly of coarse, black, volcanic cinders, almost lacking loamy admixtures. The adobe soils are derived from the decomposition of certain kinds of lava. They occur throughout the reserve, but are best developed on the slopes leading to Verde Valley in the south-west portion of the reserve. They also constitute the bottom of all the temporary lakes and ponds in the region.

The loamy soils are moderately pervious to water; the cinder and scoriaceous soils absorb water as readily as a sponge and part with it as freely. The adobe soils are not readily pervious; they quickly shed most of the water which falls upon them, except where badly cracked, when the fissures take up large quantities. However, they retain what is absorbed with great tenacity. The loamy soils are adapted to the growth of forests best of any in the reserve; next comes the scoriaceous soils with loamy admixtures. The pure cinder soils are extremely barren and sterile, while the pure adobe soils, owing to the cracking propensity of the surface, are ill adapted to forest growth.

Most of the surface is bowlder strewn or covered with small fragments of lava, or, in the vicinity of volcanic vents, with coarse slag and scoriaceous matter. On many tracts the soil is extremely thin and imperfectly covers the underlying rough lava, constituting what is here termed "scab land."

DRAINAGE.

The visible portion of the run-off, or permanent surface flow, is small, as most of the precipitation sinks either within the reserve or in the desert and semidesert tracts which border it. Oak, Beaver, and Sycamore creeks are permanent streams, as is also Smith Creek, an insignificant runlet which rises on the eastern slopes of the crater rim of San Francisco Peak and flows a distance of 1 or 1¼ miles before sinking. A portion of its waters is impounded and piped to Flagstaff, which town depends upon it for supply. It is evident that the visible run-off accounts for only a small portion of the annual precipitation, most of which undoubtedly sinks into the fissures of the lava bed and does not again appear inside the reserve nor on the contiguous tracts.

A few springs exist here and there, which are small and of little importance except to the particular locality. Natural tanks, which are hollows in the rocky beds of small creeks, are scattered over the reserve, and artificial tanks have been constructed in various localities. Most of these are small, and during the summer are exposed to the trampling and wallowing of stock and become excessively filthy. Mormon Lake and Stoneman Lake are both dry, and at present are used either for pasture or agriculture. Eight years ago both lakes were full of water, Mormon Lake having a depth of 10 to 15 feet and being plentifully stocked with fish. Several smaller lakes in the southern portion of the reserve are in a similar condition.

The normal annual precipitation which falls on the central portions of the reserve between the 6,000- and 7,800-foot levels amounts approximately to 25 inches. There are, however, great fluctuations in the amounts, the records showing as low as 7 inches per annum at Flagstaff. The precipitation rapidly diminishes in quantity as the northern, eastern, and western areas of the reserve are approached, until finally a region is reached where semidesert conditions exist. With a rise in elevation above the 7,800-foot level there is evidently an increase in the annual rate of precipitation, but no records are in existence that show the amount of rainfall. There are two rainy seasons, one in July and August, the other from October to March or April. During the prevalence of the former the rains come from the southeast and east and are accompanied by a good deal of electrical disturbance and display, and are frequently torrential in character. The effect of electric storms upon the timber of the reserve is quite noticeable; in places as high as 5 per cent of the standing trees have been struck and killed by lightning, and it is a reasonable estimate that 1 per cent of the entire timber is destroyed by this means. The fall rains begin in October and come intermittently till spring opens, changing to snow as the season advances. These storms come from the southwest and often are accompanied by violent gales.

None of the elevations in the reserve reach the permanent snow line for this latitude, but on the northeast slopes of San Francisco Peak small drifts occasionally remain through the summer.

Yellc

Red

Whi

FOREST AND WOODLAND.

List of the conifers in San Francisco Mountains Forest Reserve.

Yellow pine .... Pinus ponderosa.
Limber pine .... Pinus flexilis.
Bristle-cone pine .... Pinus aristata.
Piñon .... Pinus edulis.
Single-leaf piñon .... Pinus monophylla.
Red fir .... Pseudotsuga taxifolia.
White fir .... Abies concolor.
Arizona fir .... Abies arizonica.
Engelmann spruce .... Picea engelmanni.
One-seed juniper .... Juniperus occidentalis monosperma.
Alligator juniper .... Juniperus pachyphloea.
Arizona cypress .... Cupressus guadalupensis.

A complete list of the broad-leafed trees in not available, but is in part as follows:

List of broad-leafed trees in San Francisco Mountains Forest Reserve.

Aspen .... Populus tremuloides.
Black cottonwood .... Populus trichocarpa.
Boxelder .... Acer negundo.
Sycamore .... Platanus racemosa.
Utah oak .... Quercus utahensis.
Rocky Mountain oak .... Quercus undulata.
Ash.
Black ash.
Walnut.
Mesquite.

Among the coniferous species the yellow-pine claims first rank, constituting over 99 per cent of the merchantable timber and about 90 per cent of the total forest. It is followed by one-seed juniper and piñon. The other coniferous trees form individually but small percentages of the total growth, and are confined to more or less circumscribed tracts. Among the broad-leafed species, aspen takes first rank, but is closely followed by the oaks. The rest of the species of this class consists of isolated trees or small groups scattered on the breaks to the larger canyons.

In the classification shown on Pl. I the woodlands are tracts which carry in part—sometimes almost exclusively—species of trees which in this region never attain mill-timber dimensions. As such they comprise areas supporting pure or mixed stands of juniper, piñon, and cypress. A few tracts covered with pure stands of small aspen likewise are included in the wooded areas, and in some localities a few isolated yellow pines may be included. The forested areas, on the other hand, are defined as tracts producing trees which grow to mill-timber dimensions.

ZONES OR TYPES OF ARBORESCENT GROWTH.

The arborescent growth in the reserve falls naturally into three chief types or divisions, with one which is intermediate or transitional. These types in their altitudinal extensions, and in the species which compose them, correspond in a general way to the different ratios of precipitation which prevail over the various districts in which they are found. Yet here, as elsewhere, the soil moisture, not always closely proportioned to the annual precipitation, comes into play and limits the range of the different types or extends it into areas where otherwise they would not occur.

The lowest type of growth, as regards altitudinal limitations, is that which constitutes the woodlands, a region where semiarid climatic and soil conditions obtain. Its chief range lies between the 5,700- and 6,200-foot levels, but soil aridity due to porous and nonretentive soil constituents may extend its altitudinal range to 7,000 feet, or even 7,500 feet. It mostly forms a fringe below the yellow-pine belt to the desert, treeless areas beyond. Occasionally it occurs on the slopes and summits of cinder cones, far in among the yellow-pine belt, always in such cases as a result of soil aridity.

The zone next above that of the woodlands with their juniper and piñon stands is that of the yellow pine. It corresponds in a general way to subhumid climatic conditions. Its chief altitudinal range is between the 6,200- and the 8,500-foot levels. On the warm and dry southern and eastern slopes of San Francisco Peak the type extends to elevations of 9,200 feet, ranging into the lower limits of the humid area. The type forms 99 per cent of the forest in the reserve, and its chief coniferous tree is the yellow pine. Here and there it is mixed with small percentages of red fir, alligator juniper, aspen, and oak.

Above the yellow-pine type comes the transition type. It lies as a comparatively narrow and irregular belt between the humid areas of the zone above and the subhumid tracts of the yellow-pine forests below. Its altitudinal range lies chiefly between the 8,500- and 9,800-foot levels, dipping in some localities considerably lower, and in others rising to 10,000 feet. It occupies the middle slopes of San Francisco Peak, Kendrick Peak, and the middle and upper slopes of Sitgreaves Peak. The type is composed of red fir, white fir, and Arizona fir—the latter in small proportions only—limber pine, aspen, scattered yellow pines, and, on San Francisco Peak, a few straggling Engelmann spruce.

The upper zone of timber is formed by the subalpine type of forest, and here is characteristic of climatic conditions verging into the humid. Its altitudinal range lies between the 9,800- and the 12,400-foot levels, and therefore is confined to the summits of Kendrick Mountain and the upper slopes of San Francisco Peak. The type is composed of aspen, Engelmann spruce, bristle-cone pine, with small proportions of Arizona fir at such points along its lower levels where it overlaps the transition belt of timber. Among the different conifers which compose the type, Engelmann spruce is the dominant species. Above the belt of subalpine forest lies a narrow treeless tract generally considered as situated above timber line, and for this reason devoid of forest. If such is the case, then the timber line on San Francisco Peak runs along the 12,000-foot level, leaving the highest peak projecting nearly 800 feet above timber line.

ASPECT AND CHARACTER OF TIMBER BELTS.

The woodland growth of juniper, piñon, and cypress is mostly made up of a multitude of scattering trees, small groups, copses, and stands of medium density, the latter of which occupy areas containing 100 to 300 acres. Occasionally the stands cover much larger tracts, as in township 24 north, range 6 east, where stands occur covering 6,000 to 8,000 acres. But, as a rule, the woodland type consists of many blocks of timber everywhere interrupted by numerous tortuous lanes of bare ground, varying in width from 15 to 150 feet, and irregular tracts containing 20 to 100 acres, which are either wholly devoid of trees or carry only a few scattering individuals.

The yellow-pine type of forest consists of open, continuous stands, here and there interrupted by tracts denuded of their forests through close logging. The stands surround and inclose many areas entirely devoid of arborescent growth— so-called "parks." These parks are of varying extent, from a mere glade of 5 acres up to tracts embracing 14,000 acres, as in township 21 north, range 4 east. Surrounding the base of San Francisco Peak, Kendrick Peak, and, in a lesser degree, Sitgreaves Peak, the yellow-pine forest is mixed with aspen. The aspen occurs either in stands of pure growth, covering tracts of 5 to 250 acres, where it represents the first restocking of ancient fire glades, or the species is scattered in small bunches or as isolated trees among the yellow pines. Most of the aspen stands or mixtures occur on elevations above the 7,200-foot level. Below this altitude the aspen is largely replaced by oak as a component of the type. In the northern areas of the reserve the oak rarely forms stands of pure growth except on tracts logged to exhaustion, as in township 21 north, ranges 7 and 8 east, where the restockings are largely composed of oak. Usually the species occurs as isolated trees, or in small clumps, often as mere brush forming undergrowth. On the north slope of many of the small cinder cones situated north of the Atchison, Topeka and Santa Fe Railroad, the yellow-pine forests contain admixtures of red fir standing in small copses or as scattering trees. The quantity of this species is insignificant.

The forest which forms the transition type consists of closely stocked stands, whose density is due to great quantities of aspen of all ages, in which the coniferous species are set in small groups, thin lines, or as isolated trees. The aspen stands are especially heavy and extensive on the north side of Kendrick Peak and on the north, west, and south slopes of San Francisco Peak. On the eastern side of the peaks and on the slopes of Sitgreaves Peak the aspen groves are much thinner, and the coniferous stands of the type more compact and extensive. The aspen stands in every case represent primary restocking after exceedingly destructive fires which wiped out most of the original coniferous growth.

The subalpine type of forest at its lower elevations is formed of closely stocked stands; of thin, scattered, low, and scrubby growths at the higher altitudes. Part of the type is composed of densely stocked stands of small aspen, inclosing blocks of Engelmann spruce; part of it consists of Engelmann spruce set in pure stands, or mixed in varying proportions of aspen, Arizona fir, and bristle-cone pine. As elevation increases, the aspen and Arizona fir disappear, until we reach stands composed chiefly of bristle-cone pine, with small percentages of Engelmann spruce, or stands where the relative proportion of the two species is reversed. Here, as elsewhere in the reserve, the pure aspen stands mark the first reforestations after fires which ravaged the region during Indian occupancy.

The subalpine forest is not composed of uniform continuous stands. Everywhere it is broken by open, grass-covered tracts, rocky areas, where the soil cover is too shallow for forest growth, and large fire glades, especially on the eastern slopes of San Francisco Peak, where no reforestation of the burned-over areas has yet begun..

UNDERGROWTH AND GROUND COVER.

Only small quantities of undergrowth occur in the woodland and yellow-pine types of forest. In the former type *Cowania mexicana, Fallugia paradoxa, Berberis fremonti*, together with seedlings and small saplings of the juniper and piñon, constitute the undergrowth. In the yellow-pine type aspen and oak brush form the principal kinds of undergrowth. In the transition and subalpine types the undergrowth is mostly composed of seedling trees, wild gooseberry, and raspberry. The ground cover, when present, consists of tufted and sword-forming species of grass, and occasionally a thin layer of pine needles. The sward occurs almost exclusively in the woodland stands, and is composed of mesquite grass, *Bouteloua oligostachya*. It is rarely continuous over any considerable area, but rather forms irregular mats divided by intervals of bare ground. The tufted grasses prevailing in the yellow pine belt consist chiefly of species of *Poa, Festuca*, and *Agropyron*. Most of them are strongly rooted species, and where not much pastured, grow in large, strong tufts which sometimes become tussocks. Mixed with the grass is a thin layer of pine needles, its depth depending on the time that has passed since the last fire swept over the ground. There is very little ground cover in the

A. BURNT TIMBER ON SLOPES OF SAN FRANCISCO PEAK.

B. YELLOW PINE, PINON, AND JUNIPER.

the largest seen measuring 39 inches in diameter 14 inches from the ground. The main stem was 8 feet high, forking several times, and the total height was 25 feet.

Arizona cypress.—This is an important tree of the woodland areas in the southern portions of the reserve. Its average total height is between 12 and 15 feet, but it sometimes attains a diameter of 24 inches. The main trunk is generally hollow butted and has value only for fencing and fuel.

COMMERCIAL VALUE OF TIMBER BELTS.

The following trees have commercial values:

As fuel......................All the species.
As fencing material........Oak, aspen, yellow pine, juniper, cypress.
As mill timber.............Yellow pine, limber pine, red fir, white fir, Arizona fir, Engelmann spruce.

As regards the various types, zones, or belts of timber, the woodland type supplies only fuel and fence posts; the yellow pine, mill timber and fencing material. The transition and subalpine forests, owing to their situation on remote tracts and at considerable elevations, have not as yet been drawn upon, but could supply limited quantities of mill timber and considerable quantities of fencing and fuel.

The most valuable tree in the woodland type is the juniper, or, as it is locally called, "cedar." It is commonly cut for fence posts, especially on the Verde slopes. Owing to its habit of branching close to the ground, only a small percentage of the stand, not over 20 per cent, is fit to be cut for such purposes. The piñon is of value only for fuel.

The chief lumber tree is the yellow pine. It is extensively cut and furnishes all of the mill timber sawed, used in, and exported from the region. The aspen is cut here and there for local use by settlers living in the reserve. Its chief use is for fence posts and rails.

CAPACITY OF THE FOREST IN MILL TIMBER.

In the area of forest examined, viz, 812,500 acres, the stand of mill timber is as follows:

Stand of mill timber in forest area.

	Feet B. M.
Yellow pine	2,725,288,000
Red fir	8,436,000
White fir	5,181,000
Engelmann spruce	4,449,000
Other species	204,000
Total	2,743,558,000

This amount of standing timber gives an average of 3,377 feet B. M. per acre. These figures do not, however, give a true idea of the average capacity of the uncut forest in mill timber, for the reason that the acreage included in forested areas, as given above, comprises many tracts covered with sapling stands, and others logged in all degrees from mere cullings representing 5 to 20 per cent of the volume of mill timber on the particular tract, to logging operations where 80 to 90 per cent have been removed.

It is very evident that the yellow-pine stands, even where entirely untouched by the ax, do not carry an average crop of more than 40 per cent of the timber they are capable of producing, and that the crop in the transition and lower subalpine belts does not exceed 8 per cent of the timber producing capacity of these areas. These conditions are chiefly attributable to the numerous fires which have swept over the region within the last two hundred years, carrying with them the inevitable consequences of suppression and destruction of seedling and sapling growth.

A yellow-pine forest, as nearly pure as the one in this region, always has an open growth, but not necessarily as lightly and insufficiently stocked as is the case in this forest reserve. The open character of the yellow-pine forest is due partly to the fact that the yellow pine flourishes best when a considerable distance separates the different trees or groups of trees. In part, also, the open stands are due to the habitat of the species, which chiefly is on tracts where the average ratio of soil moisture is too low to support heavily stocked stands. That owing to a low ratio of precipitation and the existence of a loose and permeable, nonretentive forest floor, the soil-moisture ratio is low on the areas in the reserve occupied by the yellow-pine type is clear enough, but yet it is sufficiently high to support much heavier stands than the present average, as is clearly proved by the quantity of timber standing on tracts not much subjected to fires during long periods. Taking the average stand of yellow pine on the uncut and unculled areas at 4,000 feet B. M. per acre, and estimating this stand as representing 40 per cent of the ultimate capacity in mature mill timber for the entire forested portion of the reserve, there would be an average of 10,000 feet B. M. per acre as the stand of mature timber, were the forest stocked to its full capacity. The present stand of mill timber in the uncut areas varies from 500 to 13,000 feet B. M. per acre. The light stands in many cases represent tracts which were burned clear, or nearly so, one hundred or one hundred and twenty years ago, and now are stocked chiefly with sapling growths, ranging in age from 35 to 90 years. The heavily stocked stands are largely composed of standards nearly approaching even-aged growths, and occur on tracts where, from some cause, fires were excluded mostly during the first five or six decades of the sapling age of the stands.

The transition type and the lower areas of the subalpine forest are capable of producing stands of Englemann spruce and of mixed forests yielding 30,000 to 40,000 feet B. M. per acre. Such stands now exist on the west slopes of San Francisco Peak in a few small areas fully stocked with Englemann spruce. At present most of the stands of these forest types consist of mere irregular remnants after fires, carrying from 0 to 1,200 feet B. M. per acre of mill timber.

DESTRUCTION OF THE FOREST.

The chief agencies through which the forests in the reserve are being destroyed are cutting, grazing, and fire.

Cutting.—Logging operations are or have been carried on in most of the forested areas in the central portion of the reserve, which are or have been tributary to railroads. In the more northern areas and over the greater portion of the southern areas the cutting has been light and in widely separated localities. The following table exhibits the total acreage on which the forest has been culled or cut, together with the percentages, broadly stated, of such culling or logging in the different areas:

Amount of culling and logging in forested areas.

	Per cent.	Acres.
Culled	0– 5	14,270
Do	5– 15	15,160
Do	15– 35	8,055
Do	35– 60	11,070
Cut	60– 95	24,780
Do	95–100	75,510
Total...............	....	148,845

The timber cut on these tracts has been converted into tie, stull, or round mining timber and saw logs. On the tracts where the culling runs less than 15 per cent chiefly tie timber has been taken, measuring from 10 to 16 inches in diameter 2 feet from the ground, and having sufficient clear length of trunk to cut two or more ties. Much of this sort of culling took place when the Atlantic and Pacific Railroad, now the Atchison, Topeka and Santa Fe, was building through the region. In some cases the tie cutters worked as much as 10 miles back on either side of the road and cut on both even and odd numbered sections alike, if the timber proved of suitable dimensions. Of the areas culled or cut more than 15 per cent have supplied both tie and mill timber. Stull timber has come chiefly from tracts cut more than 75 per cent. It has been the custom first to cut out the timber with saw-log dimensions, then after a few years to recut the tract

B

A

for its stull timber. In the recent cuttings, however, all classes of timber having a marketable value are removed at the first cuttings, and nothing is allowed to remain but a few slender saplings of no commercial value. On most of the tracts listed as cut 95 to 100 per cent of the cutting has been of this character.

A lumber company has extended a logging railway from the town of Williams into township 20 north, range 3 east, and up to the present time has cut over 10,880 acres which will approximate 51,000,000 feet B. M. Their logging and mill plant is up-to-date in all respects, and can handle 75,000 feet B. M. daily.

Another lumber company takes the output of a small mill located in township 20 north, range 8 east. About 17 sections have been cut, estimated at 34,000,000 feet B. M., principally for ties.

Grazing.—Grazing, especially sheep herding, is ruinous to the very young forest—that is, the seedling growth—and has been so ever since stock in large and close bands were first driven in. Cattle are fond of the young aspen which spring up as the first restockage on the nonforested park lands situated at the base and on the slopes of San Francisco Mountains. They browse on the tender seedlings as well as on the sprouts and suckers sent up by older roots of this species of tree. Sheep, however, are far more destructive to the young aspen growth than cattle, appearing to prefer the tender seedling and sucker growth to the tougher and less succulent grasses. Cattle occasionally browse on the young coniferous growth, especially on the season's shoots before they have acquired a resinous flavor. As this destroys the terminal buds, the trees so browsed on grow low, stunted, and brushy. In places like the region around Leroux Spring, in township 22 north, range 6 east, where a great deal of stock collects for water, one may see thousands of yellow pines 3 to 5 feet high which thus have lost the apical and lateral terminal buds of the season's growth. Most of the damage to the forest by grazing is caused by the different bands of sheep which are herded back and forth over much or all of the reserve. The damage is not done to the large pines, but to the seedling growth while still in the cotyledon stage. The seedling growth of the yellow pine springs up in the latter part of July and the beginning of August, following close on the beginning of the first summer rains. These seedlings come from seeds shed late in the previous fall. Seedling growth may likewise start up in the early spring. There is no good reason why it should not do so, as thousands of tender yellow-pine seedlings 1 to 2 inches in height had appeared in many localities by the 15th of August. By following the drift of a band of sheep, which was passing over localities where these seedlings were at all numerous, a very accurate knowledge was obtained regarding the various modes in which such tender growths are destroyed by these animals. It was found that the destruction of seedlings on any particular tract of land ranged from 50 per cent to total after a single passage over such ground by a 2,000-head, close-bunched band of sheep. Some of the seedlings were bitten off just below the cotyledons; some were nipped in the center of the rosette of these leaves; many were simply broken off and trampled into the ground, while in other cases the entire seedling, with the seed or the husk of the seed attached to the roots, had been dug up and thrown out of the ground by the hoof of the passing sheep. As the bands of sheep are frequently driven over the same ground, day after day, during several weeks, the chance of any seedling growth surviving on such areas is exceedingly small. Comparatively little attention, if any, seems to have been given by investigators to the destruction of the yellow pine while in the cotyledon stage. The plants are then so small that the growth of weeds or grass among which they nestle renders them inconspicuous, and furthermore, their aspect at this period of their existence is so very unlike a pine tree that ordinary observers would almost invariably mistake them for some kind of herbaceous plant. It is while in the cotyledon stage that greater and more widespread injury is done to the yellow pine and coniferous growth in general, by grazing animals, than at any subsequent time. In a region grazed so closely by sheep as is this reserve the destruction of the tender cotyledon growth must be immense, and the wonder is that any seedling pines escape. Some do, however, in the worst sheep-traversed tracts, but upon seeking the reason for this it is found that the majority of the plants preserved from destruction have come from seeds which accidentally lodged under stones or among various sorts of litter which afforded a measure of protection from the destructive hoofs and teeth of the animals. After the young pines have reached the age of three or four years the danger of destruction from grazing animals is small.

Fire.—Fires have been of frequent occurrence in all portions of the reserve, and the aspect of the present forest, as regards its capacity of mill timber in the yellow-pine type and its relative composition and yield in the transition and subalpine types, has been largely determined by this element. In recent years there has been very little destruction in the southern portion of the reserve, for although in places fires have run through the timber and consumed the dry grasses and brush, they have not been of sufficient intensity to ignite the standing timber.

The fires which burned during Indian occupancy and soon after the arrival of the present settlers in the region were far more widespread and destructive than those of recent years. As the yellow-pine forest is by far the most resistant to fires of any type of forest in the Pacific United States, it follows that the forest of the composite type has suffered more severely than any of the others. For these reasons the higher slopes of San Francisco Peak, Kendrick Peak, and Sitgreaves Peak, which are mostly covered with this type of forest, show more extensive ravages by ancient fires than any other localities in the reserve, unless one is disposed to regard the numerous lanes throughout the woodlands and nontimbered parks in the yellow-pine belt as representing fire glades which have never restocked. This theory may be the true one, for it is a fact that the most severe and the most extensive of the fires of modern date have burned in the woodlands in township 24 north, range 9 east, where on a tract, or on several connecting tracts aggregating 4,500 acres, the juniper and piñon stands have been completely burned out, creating just such nontimbered lanes and "parks" as one finds in many other localities in the reserve, the origin of which is not so clearly traceable as in the present case. The most extensive and serious of the latest fires, during Indian occupancy, in the composite type of forest have burned on the southern, western, and northern slopes of San Francisco Peak, covering altogether 18,000 acres. It took place about 100 or 110 years ago, and utterly laid waste a heavily stocked stand of Engelmann spruce and Arizona fir covering about 600 acres. The badly burned areas on which the destruction has been 60 per cent or more aggregate 6,790 acres.

Fires in the yellow-pine belt have marked with basal scars and sears 75 per cent of all the trees having standard dimensions. These sears vary from 6 inches to 12 feet in length and from 3 inches to 2 feet in superficial width. The loss of merchantable timber due to fire sears and scars with attendant rot, gum cracks, bending and twisting of the tree stems amounts, in the yellow-pine forest, to 6 per cent.

The greatest loss from fire in the yellow-pine type consists in the destruction of seedlings and sapling trees. Owing to the heavy grass growth which prevails on all areas not sheeped off, surface fires develop considerable heat and flame, and death is certain to all seedling growth in the cotyledon stage which may happen to stand in the path of the fire. At the same time a good deal of sapling growth, 5 to 15 years old, is sure to be consumed.

The origin of fires in recent years may, in part, be ascribed to carelessness of sheep herders, in part to sparks from the engines on the Atchison, Topeka and Santa Fe Railroad. The region is not good hunting or camping ground and few fires originate from the camps of hunting parties. But by far the larger number of fires are due to lightning strokes, and this cause has, of course, always operated. Electric storms are very numerous in this region during July and August. It is highly probable that not one passes over the reserve area without doing more or less damage to the forest through the medium of lightning stroke. Trees struck by lightning are everywhere met with, but are especially numerous north of the Atchison, Topeka and Santa Fe Railroad, and particularly in the blocks of forest growing on the cinder flats and slopes east of San Francisco Peak. Sections exist here on which 50 per cent of the mature yellow pine is either wholly or in part killed by lightning strokes. While most of the thunder showers are accompanied by rain, some are not, and when a tree is struck by lightning during a storm of this sort, a fire of more or less severity is sure to follow.

REPRODUCTION.

Reproduction of the yellow pine is, generally, extremely deficient as regards seedling and young sapling growth, except in an area lying east of Stoneman Lake and south of Mormon Lake. Apparently there has been an almost complete cessation of reproduction over very large areas during the past twenty or twenty-five years, and there is no evidence that previous to that time it was at any period very exuberant. The low reproductive ratio is due to various causes, some depending on the operation of natural agencies, others on human intervention. Those originating in natural agencies manifest themselves chiefly in deficient seed production, a factor which strikes at the very root of the existence of the species; those due to the acts of man depend on the destruction of the individual trees.

The yellow-pine forest in the reserve is, broadly speaking, a forest long since past its prime and now in a state of decadence. It is not meant to imply by this statement that forests will not grow on this part of the Arizona Plateau after the present one is gone, but that owing to climatic conditions which are slowly approaching aridity, the yellow pine is here gradually diminishing in reproductive vigor, and the ultimate result will be extinction of the species on these areas. The yellow-pine forest on this part of the central Arizona Plateau grows on an area exposed to all the effects of severe arid pressure from the east and west. From the west the aridity of the Colorado Desert is slowly creeping eastward; from the east the desert-like areas bordering Little Colorado River are extending westward. Along the north line of the reserve a broad band of semiaridity, the penumbra, as it were, of the advancing desert, already has joined these two regions and split asunder the continuity of the yellow-pine forests of northern Arizona.

The lack of reproductive vigor in the yellow pine in the reserve manifests itself in two ways, viz., first, in the paucity of cone production—that is, more specifically, only a small number of male and female flowers, or their aggregations as cones or aments, are produced on each tree, and years may pass without any such production; second, in imperfect fertilization of the ovules. The first of these two causes is due to some inherent lack of vegetative force in the tree, a cause wholly obscure to the knowledge of man. The second is entirely dependent on climatic features. It was observed that, during the time the staminate flowers of the pine were developing, hot dry winds blowing from across Colorado Desert were apt to prevail. On most of the trees these winds dried up the staminate aments before maturity of the pollen, thereby destroying its vitalizing power. As the fertilization of the ovules, in the case of the pine, is a matter largely depending on the distribution

A. ENGELMANN SPRUCE.

B. MILL SITE ON ARIZONA CENTRAL RAILROAD.

of the pollen by the wind, great quantities of the latter substance are required. Any considerable diminution of it in this respect lessens the chances for the fertilization of the ovules, and the imperfect or nonpollenization of the ovules means abortive seeds whose germinating power, if not wholly lacking, is, in any case, low. It was also noticed that occasionally, owing to these hot dry winds, the young cones with ovules pollenized or nonpollenized dried up after a few weeks of growth. This state of affairs is not a matter of recent years alone. It probably has existed for centuries. The percentage of perfect seeds produced in any one year is therefore exceedingly small; one has only to examine the cones representing different years of growth that lie under the trees to become convinced of this fact. Not over 1 per cent of their ovules have developed into perfect seeds. When it is considered that only a very small percentage of the perfect seeds stand any chance of germinating so long as the wind is the only medium of dispersion, and accidental covering the only chance for growth, and that the number of seeds capable of germination produced in any one year is exceedingly small when taken in connection with the great number of trees of cone-bearing age existing in the reserve, the reason for the low ratio of yellow-pine seedling growth in this region is patent enough.

At the higher elevations, where the transition and subalpine types of forests grow, the hot dry winds lose all or much of their intensity. They are cooler and carry more humidity. The consequence is that the production of pollen and the subsequent pollenization of the ovules are less or not at all interfered with, and abundant seed production is the result. The same thing happens on areas of the yellow pine belt situated in the lee of the higher peaks during the prevalence of the southwest and westerly gales of early summer.

The excessively close logging, which is a common practice on the private holding on the reserve, does not provide for a sufficiency of seed trees to restock the denuded areas. It was noticed that on some of the logged areas along the line of the Central Arizona Railroad the removal of the mature trees resulted in the retardation or death of the young growth. In this same district an area which had been logged clean was fired to remove the litter, and the result was also a complete destruction of the humus. On the tracts cut seedling growth generally is wholly lacking, and, aside from the natural deficiency of such growth arising from lack of seed production while the forest existed, these areas have mostly been deprived of their seed trees, making their restocking dependent on the chance seed blown in from localities where seed trees exist, or on a possible future seed production, fifty to seventy-five years hence, of the few saplings left after the cutting.

Sapling growth under 20 years of age is almost everywhere scanty. Above this age limit it is moderately abundant, and in most cases, on the uncut or moderately logged sections, sapling and pole stands varying in age from 36 to 80 years exist in sufficient numbers to restock the stands to their present volume of mill timber, or, in the case of those cut less than 80 per cent, to the volume previous to cutting. Most of the tracts cut 80 per cent and over do not contain sufficient sapling growth to restock, for the reason that where the forest has been cut to that extent much of the older sapling or maturing timber, 10 to 12 inches in basal diameter, has likewise been cut to make stull timber, and these closely logged lands will not again bear a forest equal to the one cut off during the next 220 or 250 years.

The reproduction of the aspen is everywhere, within the limits of its range, abundant. So likewise is that of the oak and of the different species forming the transition and subalpine belts of forest. The lowest reproductive ratio among the trees in these belts is that of the limber pine, closely followed by the bristle-cone pine and Arizona fir.

The species of trees characteristic of the woodland growth reproduce slowly. The seeds of the juniper appear largely to be abortive. Such is also the case, at least in some seasons, with the seeds of the piñon; besides, the seeds of this species are greatly relished by hosts of rodentia, which devour large quantities of them.

INFLUENCE OF THE FOREST ON RUN-OFF.

It is generally believed by the people of this region that the forest cover of the reserve is a potent factor in regulating the run-off from the plateau, and various irregularities in the flow of the Verde, and such of its tributaries as head on the plateau, are ascribed to the cutting of the forests there. In point of fact there is not an iota of evidence to prove that the yellow pine forest on this plateau is, in any marked degree, a factor in the regulation of the run-off from the region.

A forest to be of any importance in the retardation of the run-off from any particular region must, first of all, be thickly set; it must be of dense growth, the closer the better; it must contain a liberal amount of undergrowth, herbaceous plants, shrubs, seedling and sapling trees, and a humus layer as a cover to the soil of the forest floor. There is not a pure yellow-pine forest in the western United States that fills these indispensable conditions. So far as the water-shedding capacity of the areas within the reserve covered with yellow-pine forest is concerned, the entire forest cover could be cut away and the run-off would be as regular as at present, and no larger. The pure yellow-pine forest is always too open and scattered to afford much shade. The hot dry winds circulate freely and the evaporation is probably nearly as great as would be the case did no forest exist. Besides, the pure yellow-pine type always grows on areas where subhumid conditions bordering on semiaridity prevail, and at the best there can not possibly exist a very large quantity of water to conserve on such tracts.

A. BILL WILLIAMS MOUNTAIN.

B. FAIR STAND OF YELLOW PINE.

The real surface regulator of the run-off in this reserve is the ground cover of tufted or sward-forming grasses which hold the soil in place, split up and spread out the little rivulets of water that form during rains or while the spring break-up lasts, give time for their absorption by the soil, and thus prevent, to some extent, violent rushes of water into the channels of the different streams and runs. Much of the water absorbed by the soil finds its way into the fissures of the underlying rock, to reappear as springs in the canyons or along the breaks of the plateau. Whatever interferes with the ground cover of grass influences the character and rapidity of the run-off, because a bare surface on a slope offers no impediment to the formation of torrential currents. which soon gully and remove the soil cover. For these reasons the felling of timber on the plateau does not accelerate in any pronounced way the run-off, but the excessive grazing to which most of the reserve is subject, and which in every case sooner or later exterminates the grass cover, will and does operate to hasten the run-off, and that in no uncertain manner.

The transition and subalpine belts of timber comprise most of the features which should exist in stands of forest capable of exerting any influence on the run-off. In this reserve they occupy such a limited scope of country that, except in one instance, their beneficent work is not at all of any particular consequence. The one instance occurs on the eastern slope of San Francisco Peak, in the central crater of the mountain, now the drainage basin of Smith Creek. The watershedding capacity of this basin is of importance to the people of Flagstaff, because from Smith Creek they obtain their supply. Most of this basin was stripped of its forest by fires of 35 or 40 years ago. Very few of the recent burns are reforesting. There is no ground cover of grass, or it is exceedingly scanty, hence after a rain of any considerable magnitude the little creek for a few hours carries a rushing torrent of water, which brings down great masses of loose rock and sand. The basin previous to the fire supported a heavily stocked stand of subalpine forest, and should be permitted to restock as rapidly as possible by the rigid suppression of fires and absolute exclusion of grazing animals from all its portions. The water supply afforded by the creek is at times insufficient for the needs of the town. A fully stocked forest covering the basin would increase the flow by retarding and compelling absorption of the surface water during storms, and by the shade and protection it would afford against too rapid melting of the snows at the head of the creek in spring and early summer.

GRAZING VALUE OF THE RESERVE.

The areas of the northern Arizona Plateau now comprised within the reserve limits originally produced a luxuriant growth of grass. It yet is vigorous and abundant on tracts where, owing to various causes, chiefly lack of watering places for stock, excessive pasturing has not prevailed. To give an idea of the richness of the gramineous flora, as well as to show the composition of the ground cover of grass, there is appended a list of grasses collected in the northern part of the reserve. It should be added that the list probably does not represent more than 60 per cent of the number of species actually growing within the reserve.

Grasses collected in northern portion of San Francisco Mountains Forest Reserve.

[Determined by J. B. Leiberg.]

1. Agropyron pseudorepens S. & S.
2. A. violaceum (Hornem.) Vasey.
3. Agrostis hyemalis (Walt.) B. S. P.
4. A. verticillata Vill.
5. Andropogon glomeratus (Walt.) B. S. P.
6. A. hallii Hack.
7. A. hirtiflorus Kunth.
8. Aristida arizonica Vasey.
9. A. longiseta fendleriana (Steud.) S. & M.
10. Blepharoneuron tricholepis (Tow.) Nash.
11. Blepharidachne kingii (S. Wats.) Hack.
12. Bouteloua curtipendula (Michx.) Torr.
13. B. eriopoda Torr.
14. B. microstachya (Fourn.) Dewey.
15. B. oligostachya (Nutt.) Torr.
16. B. prostrata Lag.
17. Bromus richardsonii Link.
18. Chætochloa composita (H. B. K.) Scrib.
19. Deschampsia cæspitosa (L.) Beauv.
20. Elymus canadensis L.
21. Epicampes rigens Benth.
22. Eragrostis major Host.
23. E. neomexicana Vasey.
24. Festuca ovina arizonica (Vasey) Hack.
25. F. ovina pseudovina Hack.
26. F. ovina supina (Schur.) Hack.
27. Hilaria jamesii (Torr.) Benth.
28. Koeleria cristata (L.) Pers.
29. Melica parviflora (Porter) Scrib.
30. Muhlenbergia porteri Scrib.
31. M. racemosa (Michx.) B. S. P.
32. M. wrightii Vasey.
33. Munroa squarrosa (Nutt.) Torr.
34. Oryzopsis micrantha (T. & R.) Thurb.
35. Phleum alpinum L.
36. Panicum barbipulvinatum S. & W.
37. P. bulbosum H. B. K.
38. P. hallii Vasey.
39. Poa brevipaniculata S. & W.
40. P. coloradensis Vasey.
41. P. eatoni Wats.
42. Sitanion breviflorum J. G. S.
43. Sporobolus asperifolius (Nees) Thurb.
44. S. cryptandrus (Torr.) A. Gray.
45. S. cryptandrus flexuosus Thurb.
46. S. interruptus Vasey.
47. S. ramulosus Kunth.
48. Stipa comata Trin. & Rupr.
49. S. vaseyi Scrib.
50. Trisetum subspicatum (L.) Beauv.

The ground cover in the yellow-pine forest is chiefly composed of the following species:

Grasses collected in yellow-pine forest of San Francisco Mountains Forest Reserve.

1. Agropyron pseudorepens.
7. Andropogon hirtiflorus.
8. Aristida arizonica.
9. A. longiseta fendleriana.
23. Eragrostis neomexicana.
24. Festuca arizonica.
28. Koeleria cristata.
30. Muhlenbergia porteri.
37. Panicum bulbosum.
41. Poa eatoni.
42. Sitanion breviflorum.
48. Stipa comata.
49. S. vaseyi.

Among these species Nos. 7, 8, 24, 41, and 49 comprise 90 per cent of the entire cover, and supply the chief pasturage of the region. These species are strongly rooted grasses, growing in tufts, and where not pastured attain a height, at flowering time, of 16 to 24 inches. Nos. 21 and 30 occur in abundance along the margins of the various runs and in the open "parks" where water stands in the early spring. They form broad tufts, sometimes verging to a tussocky condition. Nos. 9 and 15 are a low sward-forming species, occurring in the drier areas of the yellow-pine forests, and extending into the woodlands.

In the woodlands No. 15 is the most common species. It is a sward-forming grass, never, when undisturbed, growing in tufts. It does not form a continuous sward, rather patches of sward interrupted by narrow intervals of bare ground, and is the common mesquite grass of the region. The tufted species of the woodlands comprise a few of those which grow in the yellow pine forests, such as Nos. 7, 8, 24, and 48. In the cinder fields east of San Francisco Peak, the chief one of the tufted species, and often the only one over considerable areas, is No. 6.

The ground cover in the transition and subalpine types of forest is chiefly composed of the following species:

Grasses in transition and subalpine types of forest in San Francisco Mountains Forest Reserve.

2. Agropyron violaceum.	28. Koeleria cristata.
3. Agrostis hyemalis.	35. Phleum alpinum.
17. Bromus richardsonii.	40. Poa coloradensis.
19. Deschampsia cæspitosa.	41. Poa eatoni.
25. Festuca ovina pseudovina.	50. Trisetum subspicatum.

Among these, Nos. 2, 17, 25, 40, and 41 constitute 70 per cent of the cover.

When the various grasses here enumerated are closely and persistently pastured, particularly by sheep, they die out or become much reduced in size and vegetative activity, and various changes take place in the character of the subsequent vegetation on such tracts. Thus, in the woodland the destruction of the turf-forming grass, No. 15, is followed by exuberant growths of various low desert shrubs and herbaceous Compositæ, particularly species of sunflowers. They color the landscape brilliantly yellow, but have little value for pasturage. Where the woodlands and the forest blend, or occasionally overlap, the destruction of the different tufted and sward-forming grasses is followed by thick growths of tall weedy Compositæ allied to the sunflower tribe, species of milk weed, wild tobacco, etc. In the yellow pine belt the extermination of the tufted grasses is followed by the growth of the following annual species of grass, namely: *Bouteloua prostrata* and *Sporobolus ramulosus*. The latter species is particularly abundant and invariably occupies the worn-out sheep runs to the almost complete exclusion of all other species except *Agropyron pseudorepens*. The two annual grasses named are of no value for pasturage purposes.

The grazing capacity of the reserve is, at the present time, too heavily taxed. Most of the area south of the Atchison, Topeka and Santa Fe Railroad is eaten or sheeped out, and like conditions prevail on most of the west half, as well as on the north and east tiers of townships in the area of the reserve situated north of that railroad. The region having the best grazing value is now confined to a circle of 9 or 10 miles radius, with San Francisco Peak for its center.

The destruction of the pasturage is chiefly due to excessive sheep herding, more particularly to the trampling of the herd than to grazing. After July 1 sheep eat comparatively little of the larger and coarser grasses. They browse extensively on weeds, young aspen when obtainable, and on some of the more tender sprouts or stalks of the less conspicuous sorts of grass; but the trampling continues, and on a run where any considerable number of sheep have been herded for three or four weeks the strongest and most vigorous growth looks as though some mighty leveling or crushing machine had passed over it.

EFFECTS OF SHEEP HERDING ON FOREST FLOOR.

The immediate effect of excessive sheep herding is invariably to loosen and pulverize the soil to a dust-like consistency. The pulverizing effect may extend to a depth of 2 to 4 inches during a season, depending on the number of bands passing over any given area and the length of time they are confined to or herded on any particular run. The loosening of the top soil facilitates its wasting and washing by rains and running water, and as several successive loosenings and washings may occur during a season on the same area, the amount of soil removed may be considerable. The evil effects that follow the breaking up of the surface of the forest floor are much intensified where the sward of grass or the tufted species of this class of vegetation have been cropped off clean, and the wastage of the soil by running water is thus greatly accelerated on such areas. The character of the slope, whether steep or of easy gradient, is an important factor. In the portion of the reserve north of the Atchison, Topeka and Santa Fe Railroad the slope is slight and in many places rather indefinite; hence, with the exception of the immediate slopes of San Francisco Peak, Kendrick Peak, and Sitgreaves Peak the loosening of the soil, so far as it affects its wastage into the different runs and stream channels, is of little consequence. But on the areas south of that line of railroad the reserve abounds in ravines and canyons with steep rocky slopes. The sheeping of these slopes, with attendant destruction of the grass cover and loosening of the soil, is productive of a steady, if not rapid denudation of these areas down to the underlying bed rock. It is true there can never be any very deep gullying in this region, because the soil cover is too thin, and the underlying hard lava does not wear to any appreciable extent in the course of centuries; but the productive soil is being removed, the barren bed rock is coming to the surface, and the lower areas of the Verde are being more and more subjected to sudden torrential floods from the

A. YELLOW PINE ON THE VERDE SLOPE.

B. BAND OF 2,200 SHEEP.

canyons heading in the plateau. This latter state of affairs is not brought about by the cutting of the timber on the plateau, but is due to the destruction of the grass cover and to the denudation of the underlying bed rock through the effects following the excessive sheep herding and other forms of close pasturing in the south half of the reserve.

NONTIMBERED TRACTS.

The nontimbered lands are such in part owing to aridity of soil and climate, in part because the lands are too rocky and the soil cover too thin to support forest; others—small tracts only—are situated at limits above arboreal growth; other tracts possess an exceptionally sterile soil, while a considerable portion of these timberless lands consist of the so-called "parks." Some of the nontimbered tracts owe their condition to deforestation by fire and deficient restockage, and a few consist of bottoms of intermittent ponds and lakelets. Aridity of soil and climate are, however, responsible for the largest acreage of timberless land. The most extensive areas are in the townships bordering the north, east, and west boundaries of the reserve. Large "parks" occur in townships 21 and 22 north, ranges 4 and 5 east, one of them containing 16,000 acres. Smaller tracts of nontimbered land are scattered throughout the reserve, chiefly as arid slopes and summits of cinder cones and ridges. Usually the timberless lands are grassed over, or, where sheeped out or otherwise too closely pastured, bear more or less close or open growths of low-growing desert or semidesert shrubs; but much of the nontimbered cinder flats and slopes of San Francisco Peak bear no vegetation of any sort. Some of the nontimbered areas are capable of reforestations, and large tracts will in time restock with forest, provided fires and grazing are reduced to a minimum; but there still remains a large percentage of these lands too arid for the growth of forest-forming species of trees, or even of the woodland type.

AGRICULTURAL VALUE.

The reserve is not an agricultural region, the area under tillage aggregating only about 2,500 acres. The lands under tillage are situated chiefly in the dry beds of Stoneman and Mormon lakes, along Oak Creek, in hollows in the lava sheet, or at the foot of ridges where local areas of seepage exist. The crops produced consist of oats, wheat, and potatoes. Fruits are not raised, not even any of the bush fruits.

DETAILED DESCRIPTIONS.

TOWNSHIP 15 NORTH, RANGE 6 EAST.

Sections 1 to 18, inclusive, are within the reserve, and are situated on the lower terraces of Colorado Plateau drained by Verde River. The soil is mainly black adobe. The woodland growth consists chiefly of one-seed juniper and single-leaf piñon, which are mostly small, although the juniper reaches fence-post dimensions, for which purpose it is frequently cut. Reproduction is not sufficient to maintain the present density of stand nor is there any advance over the nontimbered areas.

In the south half of this township, outside of this reserve, there are divers small areas of Beaver Creek bottom lands under cultivation, aggregating about 60 acres. The township has little value for pasturage, as it was long ago thoroughly sheeped. There is no timber.

TOWNSHIP 15 NORTH, RANGE 7 EAST.

Sections 1 to 24, inclusive, are within the reserve and are drained by Beaver Creek, a tributary of Verde River. The surface is steep, rough, and broken, lying on the upper terraces of Colorado Plateau. The soil is adobe with considerable stone.

There is no timber.

Stand of species in T. 15 N., R. 7 E.

	Cords.
Arizona cypress	2,300

TOWNSHIP 15 NORTH, RANGE 8 EAST.

Sections 1 to 24, inclusive, are within the reserve. The surface is steep, rolling, and broken, with a stony adobe soil. The drainage is westward into Beaver Creek.

Stand of species in T. 15 N., R. 8 E.

	Cords.	Feet B. M.
Yellow pine		6,500,000
Arizona cypress	3,820	
Oak	700	
Total	4,520	

Forest conditions in T. 15 N., R. 8 E.

Average total height of yellow pine	feet	75
Average height, clear	do	8
Average diameter, breast-high	inches	16
Dead	per cent	5
Diseased	do	20
Average age	years	160
Reproduction		Poor.

TOWNSHIP 15 NORTH, RANGE 9 EAST.

Sections 1 to 18, inclusive, are within the reserve. It is a rolling and broken area situated on the broad divide between Little Colorado and Verde rivers. The yellow pine is very scattering and of poor quality. It will log best to the south and west.

Stand of species in T. 15 N., R. 9 E.

	Cords.	Feet B. M.
Yellow pine	.....	17,000,000
Oak	480	

Forest conditions in T. 15 N., R. 9 E.

Average total height of yellow pine	feet..	75
Average height, clear	do...	10
Average diameter, breast-high	inches..	15
Dead	per cent..	2
Diseased	do...	15
Average age	years..	150
Reproduction		Medium.

TOWNSHIP 15 NORTH, RANGE 10 EAST.

Only sections 1 to 6, inclusive, of this township are within the reserve. The tract is situated on a high plateau with a gently rolling or sloping surface which drains into Little Colorado River through Sunset Canyon, which is outside the reserve boundary. The soil is stony adobe and supports a scattering forest and woodland of poor trees.

Stand of species in T. 15 N., R. 10 E.

	Cords.	Feet B. M.
Yellow pine	.....	4,000,000
Arizona cypress	235	
Oak	850	
Total	1,085	

Forest conditions in T. 15 N., R. 10 E.

Average total height of yellow pine	feet..	75
Average height, clear	do...	8
Average diameter, breast-high	inches..	16
Dead	per cent..	3
Diseased	do...	15
Average age	years..	160
Reproduction		Poor.

TOWNSHIP 15 NORTH, RANGE 11 EAST.

Only sections 1 to 6, inclusive, are within the reserve. The tract has a gently rolling surface which drains into Sunset Canyon. The soil is a stony adobe which supports only a very light scattering woodland.

There is no timber.

Stand of species in T. 15 N., R. 11 E.

	Cords.
Arizona cypress	900

TOWNSHIP 16 NORTH, RANGE 5 EAST.

The northern area consists of rolling plateau land capped with lava. In the central portion the plateau breaks off to the valley of the Verde, forming a series of magnificent cliffs 1,000 to 2,000 feet in height. The soil is gravelly loam, here and there changing to adobe. In the southern part of the township the soil is mixed with limestone débris from the cliffs.

There is no permanent flow of water except where Oak Creek crosses the northwest corner of the township. The northern tier of sections is wooded with single-leaf piñon and one-seed juniper. The growth is thin and scattering, and the trees are low, limby, and knotty, and grow mostly in places that are inaccessible. There is no reproduction. The accessible portions of this township were long since over-sheeped and the grasses have not yet recovered.

There is no timber.

TOWNSHIP 16 NORTH, RANGE 6 EAST.

In the northern portion are steep terraces where the Colorado Plateau breaks into Verde Valley. The southern portion is rolling land, cut by numerous gullies and ravines, and intersected by an intermittent tributary of Beaver Creek. The soil is loamy adobe with a stony and bowlder-strewn surface. Large areas bordering Beaver Canyon have no soil, a rough vesicular lava forming the surface. There is no permanent surface flow of water. The woodlands are chiefly in the southwest quarter and consist of one-seed juniper, single-leaf piñon, and a few yellow pine and oak, all of which are short, stunted, limby, and knotty, and have only a fuel value.

Reproduction of juniper is extremely slow and uncertain, and the small quantity of yellow pine is not increasing, nor does there appear to be any advance over the nontimbered areas. The lands, wherever accessible, have been sheeped so long that the grass has been exterminated.

There is no timber.

TOWNSHIP 16 NORTH, RANGE 7 EAST.

The northern half of this township consists of sloping lava terraces constituting the upper breaks of Colorado Plateau to Verde Valley. The southern half is rolling land; in the eastern sections the surface rises into ridges and cones of volcanic origin. It is cut by rocky runs, ravines, and gorges tributary to Beaver Creek. The soil is loamy adobe, cindery, and scoriaceous near the cones and volcanic ridges, and the entire surface is stony and bowlder strewn.

There are no perpetual streams or springs, but in section 7, in a rocky ravine, are three small natural tanks—mere hollows in the rocky bed of the stream—

which serve as catch basins for limited quantities of storm water which they hold throughout the year. The woodland is chiefly in the south half and bears one-seed juniper and single-leaf piñon. The growth is generally low, but the juniper attains sufficient size to yield fence posts, for which purpose it is occasionally used. There is but little reproduction, and the tendency is seemingly toward extinction of the species. The township has no pasturage value, having long since been sheeped out.

There is no timber.

TOWNSHIP 16 NORTH, RANGE 8 EAST.

The surface is rolling and broken. The axis of a small secondary divide crosses the township from southeast to northwest and results in some erratic drainage, which, however, is all into Verde River via the tributaries of Beaver Creek. The soil is stony adobe. The area known as Stoneman Lake is now dry and a portion of it is farmed. It is bounded on the north, east, and south by precipitous walls 500 to 600 feet high, and on the west the rim is 100 feet in height. As late as 1894 this lake contained water.

Stand of species in T. 16 N., R. 8 E.

	Cords.	Feet B. M.
Yellow pine	.....	34,000,000
Arizona cypress	2,670	
Oak	1,140	
Total	3,810	

Forest conditions in T. 13 N. R. 8 E.

Average total height of yellow pine ....feet..	80
Average height, clear ....do....	10
Average diameter, breast-high.................................inches..	16
Dead ....per cent..	5
Diseased ....do....	20
Average age...years..	165
Reproduction, in northeastern portion	Good.
Reproduction, elsewhere..	Medium.

TOWNSHIP 16 NORTH, RANGE 9 EAST.

This township lies upon the main divide between Little Colorado and Verde rivers, and although presenting no features of high relief has a rolling and broken surface, which reaches an elevation of 8,000 feet above the sea. The soil is a stony adobe which supports an old-growth forest of good-quality timber along the creek beds and ravines. On the higher elevations the timber is rough and of poorer quality. Pine Spring is located in section 13.

Stand of species in T. 16 N., R. 9 E.

	Cords.	Feet B. M.
Yellow pine	......	66,500,000
Oak	10,620	

Forest conditions in T. 16 N., R. 9 E.

Average total height of yellow pine	feet..	90
Average height, clear	do....	10
Average diameter, breast-high	inches..	20
Dead	per cent..	2
Diseased	do....	15
Average age	years..	180
Reproduction		Medium.

TOWNSHIP 16 NORTH, RANGE 10 EAST.

This township has a rolling surface with a gentle slope toward the east, draining into the Little Colorado River through Sunset Canyon. The stony adobe soil supports a fair stand of yellow pine, of which only 30 per cent is good quality, but which can be logged to the north or east very cheaply.

Stand of species in T. 16 N., R. 10 E.

	Cords.	Feet B. M.
Yellow pine	......	12,250,000
Arizona cypress	16,200	
Oak	4,270	
Total	20,470	

Forest conditions in T. 16 N., R. 10 E.

Average total height of yellow pine	feet..	80
Average height, clear	do....	8
Average diameter, breast-high	inches..	18
Dead	per cent..	5
Diseased	do....	20
Average age	years..	175
Reproduction		Poor.

TOWNSHIP 16 NORTH, RANGE 11 EAST.

The general surface is rolling, cut by two deep gulches of the Sunset Canyon drainage, tributary to which are numerous small ravines. The soil is adobe, mixed with a large proportion of stone, and supports a heavy and evenly distributed woodland of Arizona cypress and alligator juniper.

There is no timber.

Stand of species in T. 16 N., R. 11 E.

	Cords.
Arizona cypress	46,050
Alligator juniper	Unimportant

TOWNSHIP 17 NORTH, RANGE 4 EAST.

The western areas consist of rolling tracts of limestone and lava terraces, rising in the eastern portion into perpendicular cliffs 1,500 to 2,000 feet high, which here form in part the western breaks of the Colorado Plateau. The soil is a gravelly loam, chiefly derived from decomposition of the lava. There are no permanent surface streams, springs, or tanks. There is no wooded area, but here and there occur scattered trees of single-leaf piñon and one-seed juniper. The accessible areas have long since been sheeped off.

There is no timber.

TOWNSHIP 17 NORTH, RANGE 5 EAST.

This township lies on the western breaks of Colorado Plateau, and has a very rough and broken surface. Oak Creek, a perpetual stream, flows across the southeast quarter in a very deep rocky canyon, and a smaller tributary canyon cuts the township north and south. The surface is barren, rocky land, with a thin covering of loam or adobe in places, and supports no woodland save a few scattering Arizona cypress, alligator juniper, cottonwood, sycamore, and oak.

There is no timber.

TOWNSHIP 17 NORTH, RANGE 6 EAST.

This township is a high plateau of rolling land, deeply cut by the canyon of Oak Creek in its northwest quarter and by tributary canyons which drain southward. The surface is mostly stony, with some adobe on the plateau, and supports a scattered woodland and a heavy undergrowth along Oak Creek of oak, cottonwood, willow, and sycamore.

There is no timber.

Stand of species in T. 17 N., R. 6 E.

	Cords.
Arizona cypress	5,850
Oak	1,950
Total	7,800

TOWNSHIP 17 NORTH, RANGE 7 EAST.

This irregular township lies on the upper breaks of Colorado Plateau, and has a rolling surface cut by numerous canyons. The soil is adobe. The timber is chiefly upon the high plateau, and its natural outlet would be southward, although it can be logged northward should occasion require.

Stand of species in T. 17 N., R. 7 E.

	Cords.	Feet B. M.
Yellow pine	.....	15,500,000
Arizona cypress	3,200	
Oak	3,160	
Total	6,360	

Forest conditions in T. 17 N., R. 7 E.

Average total height of yellow pine	feet..	75
Average height, clear	do....	9
Average diameter, breast-high	inches..	17
Dead	per cent..	5
Diseased	do....	20
Average age	years..	170
Reproduction		Poor.

TOWNSHIP 17 NORTH, RANGE 8 EAST.

The surface is rolling and broken, and is cut by numerous small canyons and ravines, which form in part the flood drainage of Beaver Creek. The soil is a very stony adobe, with a slight admixture of loam. About 40 per cent of the yellow pine is of good quality.

Stand of species in T. 17 N., R. 8 E.

	Cords.	Feet B. M.
Yellow pine	......	71,000,000
Oak	12,800	
Arizona cypress	360	
Total	13,160	

Forest conditions in T. 17 N., R. 8 E.

Average total height of yellow pine	feet..	80
Average height, clear	do....	10
Average diameter, breast-high	inches..	16
Dead	per cent..	3
Diseased	do....	15
Average age	years..	175
Reproduction		Medium.

TOWNSHIP 17 NORTH, RANGE 9 EAST.

This township lies on both sides of the main divide between Little Colorado and Verde rivers, and has a steep, rough, and broken surface, and a stony adobe soil. It supports a fair stand of timber, of which about 40 per cent is of good quality.

Stand of species in T. 17 N., R. 9 E.

	Cords.	Feet B. M.
Yellow pine	.....	80,000,000
Oak	6,775	

Forest conditions in T. 17 N., R. 9 E.

Average total height of yellow pine	feet..	80
Average height, clear	do....	10
Average diameter, breast-high	inches..	16
Dead	per cent..	2
Diseased	do....	15
Average age	years..	175
Reproduction		Medium.

TOWNSHIP 17 NORTH, RANGE 10 EAST.

Nearly all of this township is situated upon a broad and nearly level portion of Colorado Plateau at the head of the Canyon Diablo drainage, and has less relief than any other township in the reserve. The surface soil is a very stony adobe.

Stand of species in T. 17 N., R. 10 E.

	Cords.	Feet B. M.
Yellow pine	-----	38, 250, 000
Oak	4, 800	
Arizona cypress	1, 785	
Total	6, 585	

Forest conditions in T. 17 N., R. 10 E.

Average total height of yellow pine	feet..	85
Average height, clear	do....	10
Average diameter, breast-high	inches..	18
Dead	per cent..	4
Diseased	do....	20
Average age	years..	180
Reproduction		Poor.

TOWNSHIP 17 NORTH, RANGE 11 EAST.

The surface is a gently sloping area at the head drainage of Canyon Diablo and has few features of relief. The soil is stony with an admixture of adobe and sand, and supports only a fair woodland, evenly distributed.

There is no timber.

Stand of species in T. 17 N., R. 11 E.

	Cords.
Arizona cypress	58, 150

TOWNSHIP 18 NORTH, RANGE 4 EAST.

The surface is rolling mesa land, along the west line breaking off to Sycamore Canyon in a series of cliffs 1,500 to 2,000 feet high. The soil is sandy and gravelly loam, in places adobe like, in others mixed with limestone débris. There are no permanent surface streams, springs, or tanks. All the accessible portions have been thoroughly sheeped.

There is no timber.

TOWNSHIP 18 NORTH, RANGE 5 EAST.

The northern and western tiers of sections are upon high mesa land, which breaks by cliffs 1,000 to 1,200 feet high into a very rough and rocky basin, which is tributary to Oak Creek. There is very little adobe in the rocky soil.

Stand of species in T. 18 N., R. 5 E.

	Cords.	Feet B. M.
Yellow pine	-----	11, 750, 000
Arizona cypress	1, 450	

Forest conditions in T. 18 N., R. 5 E.

Average total height of yellow pine	feet..	75
Average height, clear	do....	8
Average diameter, breast-high	inches..	16
Dead	per cent..	4
Diseased :	do....	15
Average age	years..	160
Reproduction		Poor.

TOWNSHIP 18 NORTH, RANGE 6 EAST.

This township is gashed by the canyon of Oak Creek, which is walled by limestone cliffs 1,000 to 1,200 feet high. The entire area is steep, rough, and broken, with a stony soil containing some adobe. Oak Creek is perennial.

Stand of species in T. 18 N., R. 6 E.

	Cords.	Feet B. M.
Yellow pine	.....	13,500,000
Oak	6,305	
Arizona cypress	3,625	
Total	9,920	

Forest conditions in T. 18 N., R. 6 E.

Average total height of yellow pine	feet..	80
Average height, clear	do....	9
Average diameter, breast-high	inches..	20
Dead	per cent..	4
Diseased	do....	10
Average age	years..	175
Reproduction		Medium.

TOWNSHIP 18 NORTH, RANGE 7 EAST.

This irregular township lies on the upper breaks of Colorado Plateau, has a rolling surface cut by numerous ravines and on the west side covers in part the upper terraces of Verde Valley. The soil is very stony adobe, and supports a fair woodland and some very good quality of timber, which can be logged northward to good advantage.

Stand of species in T. 18 N., R. 7 E.

	Cord.	Feet B. M.
Yellow pine	....	55,250,000
Oak	10,150	
Arizona cypress	990	
Total	11,140	

Forest conditions in T. 18 N., R. 7 E.

```
Average total height of yellow pine..........................................feet..  85
Average height, clear.....................................................do....  10
Average diameter, breast-high...........................................inches..  20
Dead...................................................................per cent..   2
Diseased..................................................................do....  10
Average age...........................................................years..  180
Reproduction............................................................ Medium.
```

TOWNSHIP 18 NORTH, RANGE 8 EAST.

This township is upon the Little Colorado-Verde divide, which here reaches an elevation of 8,400 feet at Mormon Mountain, in sections 2 and 11. The relief is not great, as the mountain rises only 1,000 feet above the adjoining areas of plateau, which present a gently rolling surface. The soil is a stony adobe. On the exposed elevations the timber is short and rough, but in the protected areas and along the ravines there is some very good quality yellow pine, which can be logged northward.

Stand of species in T. 18 N., R. 8 E.

	Cords.	Feet B. M.
Oak..	9,400	....
Yellow pine..		105,000,000
Red fir...		1,750,000
Total...		106,750,000

Forest conditions in T. 18 N., R. 8 E.

```
Average total height of timber trees..................................feet..  90
Average height, clear.....................................................do....  11
Average diameter, breast-high. ........................................inches..  20
Dead...................................................................per cent..   2
Diseased..................................................................do....  10
Average age....................................................... ......years..  180
Reproduction............................................................ Good.
```

TOWNSHIP 18 NORTH, RANGE 9 EAST.

There are three features of surface relief in this township. The eastern half is a nearly level plateau, which marks the summit of the divide between the drainage of Canyon Diablo and Walnut Canyon. The greater portion of the west half of the township is a depression about 200 feet deep, which until 1894 held a body of water 10 to 15 feet deep, known as Mormon Lake, but which to-day is sufficiently dry to be crossed by wagon roads and to furnish some good pasturage land. The Little Colorado-Verde divide borders the lake bed on the west, and its lower slopes furnish the only bold topography in the township. The soil is a stony adobe. About 40 per cent of the yellow pine is of good quality, and can be cheaply logged northward.

Stand of species in T. 18 N., R. 9 E.

	Cords.	Feet B. M.
Yellow pine	.....	24,250,000
Oak	3,405	
Arizona cypress	1,170	
Total	4,575	

Forest conditions in T. 18 N., R. 9 E.

Average total height of yellow pine	feet..	95
Average height, clear	do....	10
Average diameter, breast-high	inches..	22
Dead	per cent..	2
Diseased	do....	15
Average age	years..	190
Reproduction		Medium.

TOWNSHIP 18 NORTH, RANGE 10 EAST.

The surface is gently rolling excepting along the east boundary, where numerous ravines tributary to Canyon Diablo make some relief. The stony adobe soil supports a very scattering woodland and a small area of poor timber. It is easily accessible from any direction and can be logged cheaply.

Stand of species in T. 18 N., R. 10 E.

	Cords.	Feet B. M.
Yellow pine	.....	4,000,000
Arizona cypress	3,975	

Forest conditions in T. 18 N., R. 10 E.

Average total height of yellow pine	feet..	80
Average height, clear	do....	10
Average diameter, breast-high	inches..	20
Dead	per cent..	5
Diseased	do....	15
Average age	years..	180
Reproduction		Poor.

TOWNSHIP 18 NORTH, RANGE 11 EAST.

The rolling surface of this township is drained by numerous ravines into one of the principal tributaries of Canyon Diablo. The stony adobe soil supports a good, evenly distributed woodland. There is no timber.

Stand of species in T. 18 N., R. 11 E.

	Cords.
Arizona cypress	73,500
Alligator juniper	Unimportant.

TOWNSHIP 19 NORTH, RANGE 1 EAST.

The surface is rolling, with a slope to the southwest into Hall Canyon, a tributary of Verde River. The soil is an adobe, much mixed with stone, and supports a very thin and scattering woodland. There is no timber.

Stand of species in T. 19 N., R. 1 E.

	Cords.
Arizona cypress	3,810

TOWNSHIP 19 NORTH, RANGE 2 EAST.

The upper terraces of Colorado Plateau cross this township, resulting in a very steep, rolling, and broken surface, with a rocky soil containing some adobe. The woodland is somewhat evenly distributed, but the timber is all above the brink and only about 30 per cent is of good quality.

Stand of species in T. 19 N., R. 2 E.

	Cords.	Feet B. M.
Yellow pine	......	13,500,000
Arizona cypress	19,250	
Oak	400	
Total	19,650	

Forest conditions in T. 19 N., R. 2 E.

Average total height of yellow pine........feet..	90
Average height, clear........do....	10
Average diameter, breast-high........inches..	20
Dead........per cent..	2
Diseased........do....	10
Average age........years..	200
Reproduction	Medium.

TOWNSHIP 19 NORTH, RANGE 3 EAST.

The greater portion of this township is situated on a broad divide between Hall Canyon and Sycamore Canyon, and has a very steep, rough, and broken surface which will make the timber difficult and expensive to handle. Above the upper terraces of Verde Valley the yellow pine is evenly distributed, and about 50 per cent is of good quality.

Stand of species in T. 19 N., R. 3 E.

	Cords.	Feet B. M.
Yellow pine	......	81,500,000
Oak	7,500	
Arizona cypress	3,675	
Total	11,175	

Forest conditions in T. 19 N., R. 3 E.

Average total height of yellow pine	feet	80
Average height, clear	do	10
Average diameter, breast-high	inches	17
Dead	per cent	3
Diseased	do	20
Average age	years	190
Reproduction		Medium.

TOWNSHIP 19 NORTH, RANGE 4 EAST.

This township is divided by Sycamore Canyon which has a depth of from 1,000 to 1,500 feet. All the surface is steep, rolling, and broken and will make the timber difficult and expensive to handle. The yellow pine is exceptionally fine for this district, although not large nor clear on the average. The soil is stony with some adobe.

Stand of species in T. 19 N., R. 4 E.

	Cords.	Feet B. M.
Yellow pine		54, 250, 000
Oak	9, 420	
Arizona cypress	550	
Total	9, 970	

Forest conditions in T. 19 N., R. 4 E.

Average total height of yellow pine	feet	80
Average height, clear	do	10
Average diameter, breast-high	inches	17
Dead	per cent	2
Diseased	do	20
Average age	years	170
Reproduction		Medium.

TOWNSHIP 19 NORTH, RANGE 5 EAST.

This township lies on a broad flat divide between Sycamore Canyon and Oak Creek. The surface is rolling and gently sloping with a stony adobe soil. About 50 per cent of the timber is of good quality and will log northward to the best advantage.

Stand of species in T. 19 N., R. 5 E.

	Cords.	Feet B. M.
Yellow pine		85, 750, 000
Oak	4, 925	
Arizona cypress	2, 800	
Total	7, 725	

Forest conditions in T. 19 N., R. 5 E.

Average total height of yellow pine	feet..	85
Average height, clear	do....	14
Average diameter, breast-high	inches..	20
Dead	per cent..	2
Diseased	do....	18
Average age	years..	200
Reproduction		Medium.

TOWNSHIP 19 NORTH, RANGE 6 EAST.

This irregular township covers a large and very irregular basin at the head of Oak Creek and is a rough and broken area. With the exception of a small tract along the east line of the township, it will be very difficult and expensive to log. About 40 per cent of the yellow pine is of good quality. Along the creek is a dense undergrowth of oak, sycamore, cottonwood, willow, and pine.

Stand of species in T. 19 N., R. 6 E.

	Cords.	Feet B. M.
Yellow pine	......	78,750,000
Oak	21,400	

Forest conditions in T. 19 N., R. 6 E.

Average total height of yellow pine	feet..	80
Average height, clear	do....	10
Average diameter, breast-high	inches..	17
Dead	per cent..	2
Diseased	do....	17
Average age	years..	165
Reproduction		Medium.

TOWNSHIP 19 NORTH, RANGE 7 EAST.

This irregular township straddles the main divide between Little Colorado and Verde rivers and has a rolling and gently sloping surface. About 50 per cent of the yellow pine is of good quality and can be logged very cheaply northward.

Stand of species in T. 19 N., R. 7 E.

	Cords.	Feet B. M.
Yellow pine	.....	91,250,000
Oak	5,395	

Forest conditions in T. 19 N., R. 7 E.

Average total height of yellow pine	feet..	85
Average height, clear	do....	11
Average diameter, breast high	inches..	18
Dead	per cent..	2
Diseased	do....	15
Average age	years..	180
Reproduction		Medium.

TOWNSHIP 19 NORTH, RANGE 8 EAST.

The rolling and broken surface of this township is drained by Walnut Canyon and supports an evenly distributed stand of timber of which about 40 per cent is of good quality. A logging railroad has been operated from the north as far as section 8 with a sawmill in T. 20 N., R. 8 E.

Stand of species in T. 19 N., R. 8 E.

	Cords.	Feet B. M.
Yellow pine	.....	56,000,000
Oak	1,580	

Forest conditions in T. 19 N., R. 8 E.

Average total height of yellow pine ....feet..	80
Average height, clear..do....	11
Average diameter, breast-high...inches..	18
Dead...per cent..	5
Diseased...do....	20
Average age..years..	180
Reproduction .... Medium.	

TOWNSHIP 19 NORTH, RANGE 9 EAST.

The surface of this township is gently rolling, except some of the terraces of the Little Colorado drainage, which are steep. The timber is mostly old growth and about 40 per cent is of good quality and easily accessible. A logging railroad was built into this township, but has been abandoned and the rails removed.

Stand of species in T. 19 N., R. 9 E.

	Cords.	Feet B. M.
Yellow pine	......	23,500,000
Arizona cypress	12,850	
Oak	3,545	
Total	16,395	

Forest conditions in T. 19 N., R. 9 E.

Average total height of yellow pine ....feet..	90
Average height, clear..do....	11
Average diameter, breast-high...inches..	18
Dead...per cent..	5
Diseased...do....	15
Average age..years..	185
Reproduction .... Poor.	

TOWNSHIP 19 NORTH, RANGE 10 EAST.

The southwestern portion of this township is nearly level plateau land which breaks by a steep terrace into the slopes of the Canyon Diablo tributaries. The soil is a stony adobe.

There is no timber—only a few scattering yellow pines.

Stand of species in T. 19 N., R. 10 E.

	Cords.
Arizona cypress	40,875
Oak	500
Total	41,375

TOWNSHIP 20 NORTH, RANGE 1 EAST.

This township contains the northern end of the upper terraces of Verde Valley and presents a rolling and broken surface, except in the northeast portion, which is comparatively unbroken. The soil is a stony adobe. Both woodland and forest show a rather poor growth and the timber will be difficult and expensive to handle.

Stand of species in T. 20 N., R. 1 E.

	Cords.	Feet B. M.
Yellow pine	.....	9,250,000
Arizona cypress	2,425	
Oak	910	
Total	3,335	

Forest conditions in T. 20 N., R. 1 E.

Average total height of yellow pine	feet..	80
Average height, clear	do....	8
Average diameter, breast-high	inches..	18
Dead	per cent..	5
Diseased	do....	20
Average age	years..	170
Reproduction		Poor.

TOWNSHIP 20 NORTH, RANGE 2 EAST.

The surface is nearly all rough and broken, with steep terraces and numerous deep ravines. A secondary divide crosses this township and suggests that it be logged both northward and southward, although at the best it will be difficult and expensive logging. The best of the timber is along the eastern boundary and by means of a railway can be made tributary to the city of Williams.

Stand of species in T. 20 N., R. 2 E.

	Cords.	Feet B. M.
Yellow pine	....	69,000,000
Oak	5,675	
Arizona cypress	720	
Total	6,395	

Forest conditions in T. 20 N., R. 2 E.

Average total height of yellow pine	feet..	90
Average height, clear	do....	10
Average diameter, breast-high	inches..	20
Dead	per cent..	5
Diseased	do....	20
Average age	years..	190
Reproduction		Medium.

TOWNSHIP 20 NORTH, RANGE 3 EAST.

The surface is rolling, with a gentle slope toward the east, and is drained by Sycamore Canyon. The soil is adobe and very stony. The timber can be logged northward to good advantage. Some years ago a lumber company had a line of logging railroad across the northern part of this township and cut most of the timber tributary to it, but at present they are operating farther west. The timber now standing is about 40 per cent good quality.

Stand of species in T. 20 N., R. 3 E.

	Cords.	Feet B. M.
Yellow pine	......	82,000,000
Oak	6,925	

Forest conditions in T. 20 N., R. 3 E.

Average total height of yellow pine	feet..	85
Average height, clear	do....	10
Average diameter, breast-high	inches..	17
Dead	per cent..	3
Diseased	do....	22
Average age	years..	185
Reproduction		Medium.

TOWNSHIP 20 NORTH, RANGE 4 EAST.

This township is badly cut up by six deep canyons tributary to Sycamore Canyon and the high land is rough and broken. There is a good stand of timber, which is unevenly distributed and about 50 per cent of it is of good quality. It will have to be logged southward and will be difficult and expensive to handle.

Stand of species in T. 20 N., R. 4 E.

	Cords.	Feet B. M.
Yellow pine	......	97,500,000
Oak	5,555	

Forest conditions in T. 20 N., R. 4 E.

```
Average total height of yellow pine ..................................feet.. 90
Average height, clear .................................................do.... 10
Average diameter, breast-high .....................................inches.. 18
Dead ............................................................per cent.. 2
Diseased .............................................................do.... 20
Average age .......................................................years.. 190
Reproduction ................................................... Medium.
```

TOWNSHIP 20 NORTH, RANGE 5 EAST.

Areas along the north and west boundaries are upon the rough and broken slopes of Sycamore Canyon, but the rest of the township is on Colorado Plateau and has a rolling surface. The soil is adobe and stone and supports an exceptionally heavy forest of yellow pine, of which about 50 per cent is of good quality. All the timber on the plateau can be logged northward very cheaply.

Stand of species in T. 20 N., R. 5 E.

	Cords.	Feet B. M.
Yellow pine	......	151,500,000
Oak	6,885	

Forest conditions in T. 20 N., R. 5 E.

```
Average total height of yellow pine .................................feet.. 95
Average height, clear ................................................do.... 11
Average diameter, breast-high ....................................inches.. 22
Dead ...........................................................per cent.. 2
Diseased ...........................................................do.... 15
Average age ...................................................... years.. 215
Reproduction ................................................... Medium.
```

TOWNSHIP 20 NORTH, RANGE 6 EAST.

The surface is rolling, with terraces and long gentle slopes formed by successive flows of lava and cinder cones 500 to 600 feet high on sections 3 and 8. Long ridges and combs composed of lava and scoriaceous matter intersect the township in various directions. The eastern tier of sections forms a portion of a broad terrace of lava with a steep rocky front facing the east, which forms in part the divide between Oak Creek and Sycamore Canyon. The soil consists of sandy or gravelly loam. In depressions where water stands during a portion of the year the soil is black adobe. Most of the surface is covered with small stones and a portion is strewn with large bowlders of lava or scoriæ. In most localities the soil is very thin and the underlying rough lava outcrops.

There are no permanent streams. Springs of small volume break out in sections 3, 8, 13, and 24. Vernal lakes are located in sections 5, 6, 31, and 32.

Stand of timber species in T. 20 N., R. 6 E.

	Feet B. M.
Yellow pine	94,500,000

Forest conditions in T. 20 N., R. 6 E.

Average total height	feet..	85
Average height, clear	do....	10
Average diameter, breast-high	inches..	20
Diseased	per cent..	3
Average age	years..	190
Reproduction, east half of township		Good.
Reproduction, west half of township		None.

TOWNSHIP 20 NORTH, RANGE 7 EAST.

Most of this township consists of broad, gently rolling limestone terraces, interspersed with low ridges and hummocks, here and there capped with outflows of lava, and in sections 12 and 33 rising into low cinder cones and ridges of volcanic origin. Most of the soil consists of gravelly or sandy loam mixed with limestone débris. In depressions on the limestone terraces in the eastern part of the township much of the soil consists of fine, silty loam. The southern and western portions, which are more or less blanketed with outflows of lava, have a thin, in some places, adobe-like soil with a stony or bowlder-strewn surface, and constitute scab land.

There are no permanent streams, but in sections 7 and 12 are small springs, and in section 19 is a spring which furnishes a considerable quantity of water for the Atchison, Topeka and Santa Fe Railroad station at Flagstaff. Small tracts in the north half of the township possess a fair grazing value, being covered with a close growth of strongly rooted, tufted grasses, but most of the township has been closely sheeped, and its grazing value is exceedingly low.

Stand of timber species in T. 20 N., R. 7 E.

	Feet B. M.
Yellow pine	50,340,000

Forest conditions in T. 20 N., R. 7 E.

Average total height	feet..	90
Average height, clear	do....	12
Average diameter, breast-high	inches..	22
Average age	years..	210
Reproduction		Slow and uncertain.

Township 20 North, Range 8 East.

Walnut Canyon crosses this township from southeast to northwest. Along the north and west boundaries is some steep and broken land, but the greater portion of the surface is nearly level or rolling. The soil is a very stony adobe. The timber is easily accessible to the Atchison, Topeka and Santa Fe Railroad at Flagstaff by logging railway to the sawmill in section 34. The daily capacity of the mill is 25,000 feet B. M. The best of the yellow pine is in the northwest corner of the township and about 30 per cent of it is of good quality.

Stand of species in T. 20 N., R. 8 E.

	Cords.	Feet B. M.
Yellow pine	.....	24,000,000
Arizona cypress	5,700	
Oak	1,410	
Total	7,110	

Forest conditions in T. 20 N., R. 8 E.

Average total height of yellow pine.....feet..	75
Average height, clear.....do....	8
Average diameter, breast-high.....inches..	20
Dead.....per cent..	6
Diseased.....do....	20
Average age.....years..	180
Reproduction.....Poor.	

Township 20 North, Range 9 East.

The upper terraces of the broad divide between Walnut Canyon and Canyon Padre cross the southern part of this township and result in some steep and broken ground. Below the terraces the land is gently sloping. There is a heavy woodland, but the only timber is in the southwest corner.

Stand of species in T. 20 N., R. 9 E.

	Cords.	Feet B. M.
Yellow pine	......	1,000,000
Arizona cypress	65,900	
Oak	2,035	
Total	67,935	

Forest conditions in T. 20 N., R. 9 E.

Average total height of yellow pine.....feet..	75
Average height, clear.....do....	8
Average diameter, breast-high.....inches..	16
Dead.....per cent..	5
Diseased.....do....	25
Average age.....years..	180
Reproduction.....Poor.	

TOWNSHIP 20 NORTH, RANGE 10 EAST.

This township lies on the gentle slopes near the head of Canyon Padre, and supports only an evenly distributed but light woodland.

There is no timber.

Stand of species in T. 20 N., R. 10 E.

	Cords.
Arizona cypress	7,200

TOWNSHIP 21 NORTH, RANGE 1 EAST.

The surface is rolling with some steep areas in the vicinity of Bill Williams Mountain. The timber is mostly of poor quality, but is easily accessible to the Atchison, Topeka and Santa Fe Railroad, which crosses the northern portion of the township.

Stand of species in T. 21 N., R. 1 E.

	Cords.	Feet B. M.
Yellow pine	.....	8,250,000
Arizona cypress	2,833	
Oak	185	
Total	3,018	

Forest conditions in T. 21 N., R. 1 E.

Average total height of yellow pine	feet..	75
Average height, clear	do ..	8
Average diameter, breast-high	inches..	16
Dead	per cent..	4
Diseased	do....	25
Average age	years..	160
Reproduction		Poor.

TOWNSHIP 21 NORTH, RANGE 2 EAST.

This township includes within its boundaries nearly all of Bill Williams Mountain, which reaches an elevation of 9,264 feet, or about 2,500 feet above the adjoining plateaus. About 40 per cent of the standing timber is of good quality, and is accessible to the logging railroad, which is in operation in this township. The timber on the slopes of Bill Williams Mountain will be difficult and expensive to handle, as the surface is steep and rough.

Stand of species in T. 21 N., R. 2 E.

	Cords.	Feet B. M.
Yellow pine	.....	36,000,000
Oak	1,170	

Forest conditions in T. 21 N., R. 2 E.

```
Average total height of yellow pine ------------------------------feet..   85
Average height, clear -------------------------------------------do....   10
Average diameter, breast-high -----------------------------------inches..  18
Dead ----------------------------------------------------------per cent..   2
Diseased ------------------------------------------------------do....   10
Average age ---------------------------------------------------years.. 170
Reproduction -------------------------------------------------- Poor.
```

TOWNSHIP 21 NORTH, RANGE 3 EAST.

This township drains southward into Sycamore Canyon. The surface is level or gently rolling land, terraced by thin flows of lava which frequently inclose or partly surround small level areas. Cinder cones or their slopes, together with low ridges of volcanic origin, occur in sections 3, 5, 6, 8, 9, 10, 15, 16, 19, 31, and 32, and vary in height from 300 to 800 feet, with long slopes. The soil is gravelly loam, largely mixed with volcanic detritus. On several of the levels, especially such as hold water during a portion of the year, the soil is black or gray adobe. Most of the surface is stony or bowlder strewn and on many of the flats and terraces the soil is too thin to cover the underlying rough lava, and the tracts constitute scab land.

There are no permanent flowing streams, but in Sycamore Canyon, in section 35, water usually stands in communicating pools during most of the year. Springs of small volume occur in sections 14, 23, 27, and 33, while artificial tanks, made by damming small runs and ravines, exist here and there. The pasturage value of the township is low.

Stand of species in T. 21 N., R. 3 E.

```
                                                          Feet B. M.
Yellow pine ------------------------------------------- 40, 335, 000
```

Forest conditions in T. 21 N., R. 3 E.

```
Average total height ----------------------------------------feet..   85
Average height, clear ---------------------------------------do....   10
Average diameter, breast-high -------------------------------inches..  20
Average age -------------------------------------------------years.. 190
Reproduction ------------------------------------------------ Poor.
```

TOWNSHIP 21 NORTH, RANGE 4 EAST.

The larger portion of this township consists of level or gently rolling prairie, or park land, as it is locally known, margined on the west and east by ranges of low hills of volcanic origin, formerly cinder cones, having long since lost their conical form through erosion. There are a few shallow rocky runs in the east and west tiers of sections. The soil is gravelly loam, becoming scoriaceous and cindery on the hilly areas. Portions of the surface are bowlder strewn.

The drainage is to Verde River through Sycamore Canyon. There is no permanent surface flow, but a few small springs or tanks exist on section 32 in the bed of one of the heads of the canyon. In section 18 is a shallow well supplying small quantities of water. The township has been closely sheeped and the grazing value is therefore only moderate. It is pastured each year by cattle and sheep in excess of the recuperative power of the grass and a few years hence will have little or no pasturage value.

Stand of timber species in T. 21 N., R. 4 E.

	Feet B. M.
Yellow pine	39,350,000

Forest conditions in T. 21 N., R. 4 E.

Average total height	feet..	85
Average height, clear	do....	10
Average diameter, breast-high	inches..	20
Average age	years..	190
Reproduction, on sections 25, 32, 33, 34		Good.
Reproduction, elsewhere		Poor.

TOWNSHIP 21 NORTH, RANGE 5 EAST.

This township consists in part of rolling and in part of level tracts of land, the western tier of sections rising into ranges of hills of volcanic origin 300 to 800 feet high. The northern half of the township is terraced by successive flows of lava, the southern half consists of short terraces and occasional low hillocks of limestone, here and there overlain with thin blankets of lava outflows from local fissures. The different runs and gullies which cut the northern half of the township are few and shallow. In the south half they increase in number, deepen rapidly, and cut through the limestone formations with box-shaped canyons 30 to 120 feet deep. The soil in the north half of the township is a gravelly loam occasionally changing into black or gray adobe. In the south half the soil is much mixed with limestone débris and becomes a loamy gravel.

There are large springs in sections 10 and 11 and smaller springs in section 4. There are several artificial tanks, the largest of which are in sections 10 and 11, and supply water to the Atchison, Topeka and Santa Fe Railroad station at Belmont. All the drainage is into Verde River.

Stand of timber species in T. 21 N., R. 5 E.

	Feet B. M.
Yellow pine	109,520,000

Forest conditions in T. 21 N., R. 5 E.

Average total height..feet.. 90
Average height, clear...do.... 12
Average diameter, breast-high..inches.. 22
Average age ....years.. 210
Reproduction, on sections 13, 14, 16, 22, 24, 26, 27, 34........................ Poor.
Reproduction, elsewhere... Good.

TOWNSHIP 21 NORTH, RANGE 6 EAST.

The surface is chiefly rolling, with terraces formed principally by successive flows of lava, also, possibly, through elevations by faulting. Cinder cones and low hills of volcanic origin occur in sections 2, 15, 18, 20, 24, 25, and 27. Most of section 25 is covered by a cone rising 250 to 300 feet above the general level, with a central depression comprising about 150 acres. The soil is a gravelly loam, changing to cindery and scoriaceous on the slopes of the cinder cones. On the flats, wherever water stands during a portion of the year, as in sections 31 and 32, it consists of black or gray adobe. Most of the surface is exceedingly stony and bowlder strewn and constitutes scab land.

There is no permanent surface stream. Small springs occur on sections 14 and 20 and a shallow well on section 25. Rogers Lake and the depression mentioned on section 25 hold water in the early spring and then form small lakes. The township has long been closely sheeped and its grazing value is now very low or insignificant.

Stand of timber species in T. 21 N., R. 6 E.

Feet B. M.

Yellow pine 42,275,000

Forest conditions in T. 21 N., R. 6 E.

Average total height...feet.. 80
Average height, clear ....do.... 10
Average diameter, breast-high ....inches.. 18
Average age ....years.. 175
Reproduction, on sections 23, 25, 29, 32, 33 .... Good.
Reproduction, elsewhere .... Very little.

TOWNSHIP 21 NORTH, RANGE 7 EAST.

Sections 1 and 2 comprise rocky southern slopes of Elden Mesa, rising 2,000 to 2,200 feet above the adjoining levels. The remainder of the township consists of low, broad lava terraces or mesa land, alternating in the southern sections with level depressions which change to a succession of low swells of limestone outcrops separated by box canyons. The northern half of the township is cut by a few shallow runs and gullies; the south half is intersected by numerous ravines, which in sections 35 and 36 develop into deep and rocky canyons and form, in part, the head of Walnut Canyon. The soil consists of sandy or gravelly loam, in the south half much mixed with limestone débris. There is no permanent surface flow. There is a small spring on section 4. The value of the lands for pasturage is low. Sheep herding has been carried on for many years in excess of the recuperative power of the grass, and, with the exception of some of the sections in the south half of the township, weedy growths have replaced the former turf or tufted species of grass.

Stand of timber species in T. 21 N., R. 7 E.

Feet B. M.
Yellow pine 33,040,000

Forest conditions in T. 21 N., R. 7 E.

Average total height ....feet.. 80
Average height, clear ....do.... 10
Average diameter, breast high ....inches.. 18
Average age ....years.. 175
Reproduction .... Poor.

TOWNSHIP 21 NORTH, RANGE 8 EAST.

The north half is rolling land, terraced with successive flows of lava, intersected by low ridges of volcanic origin, and rising into small cinder cones in sections 1, 2, 4, 5, 9, and 11. Part of section 6 is on the rocky slope of Elden Mesa, rising 500 to 1,800 feet above the adjacent levels. The south half of the township consists of low outcropping limestone ridges and terraces, occasionally buried under outflows of lava, gashed and seamed with short, shallow runs and ravines tributary to Walnut Canyon. This canyon is 450 to 600 feet deep, with steep rocky slopes, about one-fourth of a mile wide between its brinks, and debouches eastward into nearly level areas of the Little Colorado slopes. The soil is mostly gravelly loam. In the south half it is mixed with limestone débris and in the north half with comminuted slag and scoriæ.

There is no permanent surface flow, nor are there any springs or natural tanks. The township is sheeped each season, besides being more or less pastured by cattle and horses. Its grazing value is yet moderate, but will not last many years longer. Some of the northern sections have been sheeped out, and now chiefly produce coarse weeds.

Stand of timber species in T. 21 N., R. 8 E.

Feet B. M.
Yellow pine 42,778.000

Forest conditions in T. 21 N., R. 8 E.

Average total height ....feet.. 80
Average height, clear ....do.... 10
Average diameter, breast high ....inches.. 18
Average age ....years.. 175
Reproduction .... Poor.

TOWNSHIP 21 NORTH, RANGE 9 EAST.

The surface is mostly level or gently rolling, with low terraces, mounds, and short ridges, all of volcanic origin, in the northern tier of sections. Walnut Canyon traverses the township from southwest to northeast, and is from 400 to 500 feet deep, with steep rocky slopes. In the northern tiers of sections the soil consists chiefly of black volcanic cinders, with small loamy admixtures. Elsewhere it is gravelly loam, with a stony or bowlder-strewn surface.

There are no permanent surface streams, springs, or tanks. Water flows through Walnut Canyon only after heavy rains, or while the spring break-up lasts. The woodlands bear stands of one-seed juniper, piñon, and scattered yellow pine along the breaks of Walnut Canyon. The reproduction is slow and scanty. The grazing value is low, for the lands have been sheeped long and closely and much of the grass is either eaten or trampled out.

There is no timber.

TOWNSHIP 22 NORTH, RANGE 1 EAST.

The surface is level or rolling and supports a very light and scattering woodland and forest in the southeast corner.

Stand of species in T. 22 N., R. 1 E.

	Cords.	Feet B. M.
Yellow pine		750,000
Arizona cypress	40	

Forest conditions in T. 22 N., R. 1 E.

Average total height of yellow pine	feet..	75
Average height, clear	do....	8
Average diameter, breast high	inches..	16
Dead	per cent..	5
Diseased	do....	25
Average age	years..	160
Reproduction		Poor.

TOWNSHIP 22 NORTH, RANGE 2 EAST.

There is some steep and broken land along the south line of this township, but the rest is nearly all rolling or level. The timber is all rough and of poor quality can be logged cheaply, and is easily accessible to the large sawmill at Williams.

Stand of species in T. 22 N., R. 2 E.

	Cords.	Feet B. M.
Yellow pine		5,000,000
Oak	325	
Arizona cypress	25	
Total	350	

Forest conditions in T. 22 N., R. 2 E.

Average total height of yellow pine ---------------------------------------feet.. 70
Average height, clear --do.... 8
Average diameter, breast-high --inches.. 16
Dead ---per cent.. 5
Diseased --do.... 25
Reproduction --- Medium.

TOWNSHIP 22 NORTH, RANGE 3 EAST.

Most of the surface is level or gently rolling, broken by low terraces of lava. Cinder and slag cones and ridges of volcanic origin rise in sections 1, 2, 3, 15, 17, 18, 19, 20, 21, 27, 29, 31, 32, 33, and 34. None of them cover any large acreage, and their height is inconsiderable, rarely exceeding 350 feet. Section 28 chiefly consists of a shallow, nearly circular depression holding water in spring and early summer. The soil is gravelly loam, here and there changing to black adobe. Almost all the land in the township has an exceedingly stony and bowlder-strewn surface. There is no permanent surface stream. The present value of this township for grazing is exceedingly low, as it has been oversheeped for many years.

Stand of timber species in T. 22 N., R. 3 E.

	Feet B. M.
Yellow pine --	15,074,000

Forest conditions in T. 22 N., R. 3 E.

Average total height---feet.. 80
Average height, clear --------,----------------------------------....do.... 10
Average diameter, breast-high --inches.. 18
Average age ---years.. 175
Reproduction --- Medium.

TOWNSHIP 22 NORTH, RANGE 4 EAST.

The surface consists of broad levels alternating with rolling land, the relief being chiefly due to successive flows of lava, which here and there have built up terraces. Extinct volcanic cones and ridges, the latter formed by outpourings from various eruptive foci, occur in sections 1, 7, 8, 10, 16, 17, 20, 23, 28, 30, and 36. None rise more than 700 feet above the general level. The cones are truncated, with crateriform central or sublateral depressions at their summits. Their slopes are usually steep and their sides are littered with blocks of slag and scoriæ. The soil is gravelly loam, sometimes changing to black adobe. Most of the the surface is stony or bowlder strewn, forming scab land.

There is no permanent surface flow of water. There is a very small spring in section 24 and another in section 28. In the north half of the township is a moderate growth of grass, partly tufted species, partly mesquite sward. It is pastured by cattle and sheep, and in some localities is suffering from overgrazing. The south half has been badly overgrazed, and has little pasturage at present.

Stand of timber species in T. 22 N., R. 4 E.

	Feet B. M.
Yellow pine	19,849,000

Forest conditions in T. 22 N., R. 4 E.

Average total height	feet..	80
Average height, clear	do....	10
Average diameter, breast-high	inches..	18
Average age	years..	175
Reproduction		Medium.

TOWNSHIP 22 NORTH, RANGE 5 EAST.

The surface comprises low ridges of volcanic origin and terraced lava flows, here and there inclosing or bordering small flats. Extinct volcanic cones or slopes leading to their summits occur on sections 3, 15, 19, 20, 22, 31, and 33. They are isolated, and are from 200 to 300 feet high. The cone in section 33 is being taken by the Atchison, Topeka and Santa Fe Railroad for road ballast, for which purpose the scoriaceous material is well adapted. The soil is gravelly loam, the proportion of gravel varying; where the quantity is small it assumes the character of adobe. The surface is stony and bowlder strewn everywhere. On the tops and along the breaks of the terraces the layer of soil is often very thin and imperfectly covers the underlying rough lava, and for this reason fully 75 per cent of the township is scab land.

There are no permanent surface streams, springs, or tanks. The grazing value of the township is moderate. The level tracts have been sheeped out, and the terraced areas are grazed each season by cattle and sheep. They bear a sparse growth of tufted grasses, chiefly species of *Poa* and *Festuca*.

Stand of timber species in T. 22 N., R. 5 E.

	Feet. B. M.
Yellow pine	79,320,000

Forest conditions in T. 22 N., R. 5 E.

Average total height	feet..	85
Average height, clear	do....	10
Average diameter, breast-high	inches..	20
Average age	years..	190
Reproduction		Medium.

TOWNSHIP 22 NORTH, RANGE 6 EAST.

Most of the lands are rolling terraced lava flows, rising from 100 to 400 feet above the small levels which they inclose or border in the central portions of the township. The surface of these lava flows is extremely broken, abounding in low hummocks, ridges, shallow runs, deeper ravines, and short abrupt canyons. Sections 1, 2, and 12 are upon the foothills of San Francisco Peak, and are very rough and broken flows of lava. In section 20 is Wing Mountain, an extinct isolated volcanic cone rising 1,000 feet above the adjacent level, on which a large dish-shaped central depression marks the site of the ancient crater. Section 31 also contains a cinder cone. Sections 23, 26, and 27 are almost level prairie. The soil is sandy, gravelly loam, deep on the level areas and shallow on the rolling lava terraces, where it imperfectly covers the underlying rough lava, and is mostly scab land.

There is no permanent surface stream. There are small springs on sections 12 and 14, one of the latter being Leroux Spring, which is a famous watering place for stock. In the vicinity of Leroux Spring the lands have been overgrazed, but elsewhere the growth of strongly rooted, tufted species of grass is good.

Stand of timber species in T. 22 N., R. 6 E.

	Feet B. M.
Yellow pine	82, 416, 000

Forest conditions in T. 22 N., R. 6 E.

Average total height	feet..	85
Average height, clear	do...	10
Average diameter, breast-high	inches..	20
Average age	years..	190
Reproduction, sections 10, 11, 13, 15, 16		Good.
Reproduction, sections 4, 5, 6, 7, 8, 9, 17, 18, 19, 25		Medium.
Reproduction elsewhere		Poor.

TOWNSHIP 22 NORTH, RANGE 7 EAST.

The greater portion of the surface is very rough and mountainous, including a part of the southern slope of San Francisco Peak and a rough outlying spur known as Elden Mesa. The spurs from San Francisco Peak terminate in terraces formed by successive flows of lava which stretch to the south for a distance of 6 or 7 miles. The upper terraces are narrow and break away with steep rocky fronts, but those lower present broad, easy-sloping tracts of rolling land, cut and seamed with gullies and canyons. Elden Mesa is an ancient volcanic center, greatly differing in its structure from the mass of San Francisco Peak and the numerous cones of the region. While called Elden Mesa, it is not at all what may correctly be termed a mesa, but is in fact a group of craters now greatly eroded, which at the close of their eruptive activity were situated around a central backbone. It rises with a steep almost precipitous rocky front from the levels which adjoin it on the east, west, and south, attaining a height of 2,000 to 2,500 feet above them.

The soil is gravelly loam, derived from decomposed lava, more or less mixed with volcanic detritus. The eastern sections are covered with great bowlder and gravel deposits, perhaps due to glacial wear of the high eastern slopes of San Francisco Peak. There is no permanent surface flow. There are four small springs, two at the southern and eastern foot of Elden Mesa and two on the southern slopes of San Francisco Peak, but the volume is small and unimportant. In the southern tier of sections the grazing value is low, but in most of the remainder of the township it is excellent, consisting of strongly rooted, tufted species of *Poa* and *Festuca*. At high elevations in sections 2, 3, 4, and 5 some tracts produce no grass, and in the thickest stands of pure aspen it is nearly lacking.

Stand of timber species in T. 22 N., R. 7 E.

	Feet B. M.
Yellow pine	36, 108, 000
White fir	3, 398, 000
Red fir	2, 124, 000
Engelmann spruce	850, 000
Total	42, 480, 000

Forest conditions in T. 22 N, R. 7 E.

Average total height	feet..	80
Average height, clear	do...	8
Average diameter, breast-high	inches..	16
Average age	years..	160
Reproduction of yellow pine		Poor.
Reproduction of other species		Good.

TOWNSHIP 22 NORTH, RANGE 8 EAST.

The surface is mostly gently rolling land, the undulating features being due to intersecting combs and ridges of volcanic origin, short and low lava terraces, low hummocks of lava, and shallow runs and ravines. Cinder cones and other forms of ancient eruptive centers occur in sections 12, 15, 22, 24, 25, 26, 27, and 34. The cones are low and of small extent. Parts of sections 18, 19, 30, and 31 form the eastern slope of Elden Mesa and rise into extremely rocky, steep, and, in some places, perpendicular escarpments. In the western sections the soil is a gravelly loam, gradually changing to cinders and scoriæ until, in the eastern areas, it consists of black cinders with very slight admixtures of loam or none whatever.

There is no permanent surface flow, except two small excavations at the foot of Elden Mesa holding a few gallons each, and fed by percolations through fissures in the rocky ledges. In the southeast corner of the township are two rock basins in the bed of a small ravine, known as Turkey Tanks, which are filled during storms and hold water throughout the year. The storm waters, which at times fill the various runs and channels, never reach beyond this point.

The western half of the township is well grassed with strongly rooted, tufted species of *Poa*, *Festuca*, and *Stipa*. In the eastern half the grass consists chiefly of mesquite sward, which, on the nontimbered tracts, has been badly sheeped, and is being gradually crowded out by close growths of coarse weeds.

Stand of timber species in T. 22 N., R. 8 E.

	Feet B. M.
Yellow pine	51, 178, 000
White fir	129, 000
Red fir	128, 000
Total	51, 435, 000

Forest conditions in T. 22 N., R. 8 E.

Average total height	feet..	80
Average height, clear	do...	10
Average diameter, breast-high	inches..	18
Average age	years..	180
Reproduction	Medium.	

TOWNSHIP 22 NORTH, RANGE 9 EAST.

The entire area of this township is covered with outpourings of lava and black cinders, the result of comparatively recent volcanic energy. Nearly every section contains cinder cones and craters which once were active in ejecting these streams of lava and vast banks and mountains of cinders. Some of the cones stand isolated, as, for example, Sunset Peak, situated partly in section 6. This peak is the most splendid and imposing example of the cinder cones found in the reserve. It rises to a height of 1,200 to 1,300 feet above the adjacent plain, forms a truncated cone with steep sides, and contains two immense funnel-shaped depressions in its center 300 to 350 feet in depth. Most of the craters in the township are grouped around central ridges or combs of lava, and form short ranges of hills. The tracts intervening between the cones and hills are level, rolling, or terraced by successive lava flows. The soil consists of black cinders and scoriæ, here and there of a brick-red color, sometimes with slight loamy admixtures, but generally lacking this ingredient. There are no permanent surface streams, springs, or tanks. Only the southwest quarter of the township possesses a slight grazing value.

Stand of timber species in T. 22 N., R. 9 E.

	Feet B. M.
Yellow pine	9, 730, 000

Forest conditions in T. 22 N., R. 9 E.

Average total height	feet..	80
Average height, clear	do....	10
Average diameter, breast-high	inches..	18
Average age	years..	175
Reproduction	Medium.	

TOWNSHIP 23 NORTH, RANGE 3 EAST.

Nearly two-thirds of this township consists of level or gently rolling land. In sections 2, 3, 4, 5, 6, 7, 10, 12, 13, 26, and 27 rise low extinct volcanic cones, none over 550 feet in height. Some of these cones are isolated, others are in groups and exist as low ranges of hills. Their slopes are not generally steep and their symmetry is largely gone owing to erosion. On the lower levels the soil is gravelly loam-gumbo, or loamy material mixed with varying proportions of volcanic detritus. In most of the sections the surface is stony and bowlder strewn, forming scab land. There are no permanent surface streams, springs, or tanks.

The grazing value is comparatively small. The ground cover of tufted grasses is low, thin, and scattering, except in portions of sections 25 and 26. The grass in the nontimbered areas, originally consisting of mesquite sward, has been closely sheeped, and the ground is now covered with a growth of coarse weeds, chiefly wild sunflowers or low semidesert suffrutescent vegetation.

Stand of timber species in T. 23 N., R. 3 E.

	Feet B. M.
Yellow pine	24,135,000

Forest conditions in T. 23 N., R. 3 E.

Average total height	feet..	85
Average height, clear	do....	10
Average diameter, breast-high	inches..	20
Average age	years..	190
Reproduction, near Mount Sitgreaves		Good.
Reproduction, elsewhere		Poor.

TOWNSHIP 23 NORTH, RANGE 4 EAST.

The surface is level or moderately undulating, except about 6,000 acres in the southwest portion, which lies on Sitgreaves Peak, and also excepting extinct volcanic cones in sections 1, 2, 3, 4, 6, 7, 12, 15, 16, 21, 23, 24, 25, 26, and 27, which are generally isolated and rise to a height of 850 feet in some cases. Sitgreaves Peak appears to have been originally an uplift of limestone, through which opened small vents and fissures, especially on the southern slope, from which issued great volumes of lava. The soil, except on the cones, is a gravelly loam, inclining to sandy in the northern foothill region of Sitgreaves Peak and occasionally assuming a gumbo-like character in the northern tier of sections. The lower levels are generally stony or bowlder strewn.

There are no permanent surface streams. Springs and artificial tanks, the latter made by throwing embankments across shallow runs to intercept storm waters, occur in sections 8, 15, and 27. The grazing value is low, as most of the township has been closely grazed by sheep, horses, and cattle.

Stand of timber species in T. 23 N., R. 4 E.

	Feet B. M.
Yellow pine	65, 672, 000
Red fir	2, 394, 000
White fir	171, 000
Limber pine	171, 000
Total	68, 408, 000

Forest conditions in T. 23 N., R. 4 E.

Average total height	feet..	80
Average height, clear	do....	10
Average diameter, breast-high	inches..	18
Average age	years..	180
Reproduction, sections 4, 5, 27, 32	Poor.	
Reproduction, elsewhere	Medium.	

TOWNSHIP 23 NORTH, RANGE 5 EAST.

Sections 2, 3, 4, 5, 6, 9, 10, 11, and 15 form the western and southern declivities of Kendrick Peak, of which the summit has an elevation, in section 2, of 9,800 feet. These tracts are made up of steep, rocky slopes, short spurs, and narrow terraces of successive lava flows which commonly terminate in precipitous fronts. Sections 1, 12, 13, 14, 16, 21, 24, 25, 26, 29, 32, 35, and 36 consist of level or moderately undulating land, roughened by low ridges and hummocks of volcanic rock due to inequalities in flow of underlying lava. Sections 7, 8, 17, 18, 19, 20, 22, 23, 27, 28, 30, 31, 33, and 34 are more or less completely covered with the remains of ancient cinder cones, which rise steeply and vary in height from 250 to 900 feet, possessing a central or sublateral depression which marks the ancient crater. Shallow ravines, runs, and gullies furrow the township in various directions. On the lower levels most of the soil is gravelly loam derived from comminuted volcanic detritus and is littered with stones and bowlders.

There are no permanent surface streams or springs. The township has an excellent value for grazing, as the grass consists of close set, strongly rooted, tufted species of the genera *Poa* and *Festuca*. It is not much grazed except in the northern sections, owing to lack of watering places. Section 22 and the adjoining ones are more closely sheeped, an artificial tank in section 22 affording a moderate amount of stock water.

Stand of timber species in T. 23 N., R. 5 E.

	Feet B. M.
Yellow pine	74, 729, 000
Engelmann spruce	1, 528, 000
Red fir	100, 000
White fir	40, 000
Limber pine	13, 000
Total	76, 410, 000

Forest conditions in T. 23 N., R. 5 E.

Average total height	feet..	80
Average height, clear	do....	10
Average diameter, breast-high	inches..	18
Average age	years..	180
Reproduction, sections 10 and 11		Good.
Reproduction, sections 9, 17, 23, 24, 25, and 36		Medium.
Reproduction elsewhere		Poor.

TOWNSHIP 23 NORTH, RANGE 6 EAST.

From the foot of San Francisco Peak this township stretches away in rolling lands sloping partly to the west, partly to the north, terraced with successive lava flows, intersected by low ranges of volcanic hills, and dotted with extinct truncated cinder cones, the more prominent of which occur in sections 2, 4, 10, 11, 12, 13, 14, 15, 16, 18, 19, 21, 22, and 27. The cones vary in height from 250 to 950 feet, usually with steep sides and with central or sublateral crateriform depressions. Most of them stand isolated but some are in groups, as in sections 15 and 16. On the lower levels the soil is gravelly loam, frequently rich and deep, especially in the swales which have for centuries received the washings from adjacent hills. In most of the township the surface is stony and bowlder strewn; particularly on the declivities of San Francisco Peak, where glaciation may have assisted in piling up the masses of bowlders which there litter the slopes.

There are no permanent surface streams, but there are a number of small springs in sections 14, 22, and 26, and tanks and wells supplying considerable quantities of water exist in sections 11, 14, 22, 23, 26, and 36. The grazing value of the township is good, and large numbers of cattle and horses are here pastured, while sheep are occasionally herded on the slopes of San Francisco Peak. The ground cover consists of a close growth of strongly rooted, tufted grasses, chiefly species of *Poa* and *Festuca*, and will afford good grazing for many years to come, provided sufficient rains fall during the growing season and the region is not overstocked.

Stand of timber species in T. 23 N., R. 6 E.

	Feet B. M.
Yellow pine	77,794,000
Red fir	391,000
Total	78,185,000

Forest conditions in T. 23 N., R. 6 E.

Average total height	feet..	80
Average height, clear	do....	10
Average diameter, breast-high	inches..	18
Average age	years..	180
Reproduction of yellow pine		Poor.
Reproduction of other species		Good.

TOWNSHIP 23 NORTH, RANGE 7 EAST.

This township consists of the main mass of San Francisco Peak, together with the larger area of its foothill region, and consequently possesses features of high relief. The surface of the foothill region is rolling and broken, being formed by a series of terraces developed by successive lava flows which issued from the slopes and base of the former great central cone. The terraces are intersected by occasional ridges and spurs of lava, raised above the surrounding levels a few hundred feet, and here and there bear a few low circular cones. Their fronts are steep and rocky, and they are seamed and gashed by numerous narrow canyons, ravines, and gullies heading in the steep slopes of the central mass of the mountain. The soil of the foothill region is composed of gravelly loam, the gravelly ingredients consisting of volcanic débris, cinders, scoriæ, fragments of lava, and the like. It is mostly thin, and over large areas imperfectly covers the underlying rough lava, hence the greater portion of the foothill region consists of scab land. The intermediate slopes of the mountains are covered with soil similar in character to that of the foothills at their lower elevations. At the higher altitudes and at the summits, the surface is slag, scoriæ, and cinders. In every section in the township bowlders and smaller fragments of rock litter the surface.

Smith Creek is a small but permanent stream heading in springs situated in sections 27 and 28. Normally the volume of its surface flow is insignificant, but during heavy rains, or when the spring break-up occurs, its volume becomes torrential. Its total length at a maximum stage is between 5 and 6 miles, when it reaches level area and sinks. During the greater portion of the year its length is not over 1 mile. This stream is of great importance because it supplies water consumed in the town of Flagstaff. Most of its normal surface flow, and some of its overflow, is intercepted in section 27, and thence piped down to the city, a distance of 15 miles. Owing to lack of proper dams to impound and store the flow when at its maximum, most of the water carried by the stream during its torrential stages is lost, as is also much of the subflow.

The foothill areas and portions of sections 22, 23, 26, 27, 28, 34, and 35, situated on the high slopes of San Francisco Peak, afford good grazing. They are pastured each season by cattle and sheep. The grass consists of strongly rooted, tufted species of *Poa*, *Festuca*, *Agropyron*, and *Stipa*.

Stand of timber species in T. 23 N., R. 7 E.

	Feet B. M.
Yellow pine	39, 827, 000
Engelmann spruce	1, 531, 000
White fir	1, 443, 000
Red fir	919, 000
Total	43, 720, 000

Forest conditions in T. 23 N., R. 7 E.

Average total height ....feet.. 80
Average height, clear ....do.... 8
Average diameter, breast-high...inches.. 16
Average age ....years.. 165
Reproduction .... Medium.

TOWNSHIP 23 NORTH, RANGE 8 EAST.

The surface is exceedingly irregular, although most of it does not present features of high relief. The most conspicuous elevation is O'Leary Peak, a mass of ancient lava rising about 2,300 feet above the surrounding plain. This peak is flanked on the south, east, and west by cinder cones of varying magnitude which present, with the exception of those in township 22 north, range 8 east, the most remarkable exhibition of volcanic action to be found within the reserve. From O'Leary Peak the township stretches to the northwest as moderately rolling ground, while to the southwest it rises into terraces originating in the lava flows from San Francisco Peak. On all the eastern and central portions of the township the soil is volcanic débris, chiefly red and black cinders. Here and there, particularly in the central areas, the cinders contain small quantities of loamy matter. In the western areas the soil contains more loam and less cinders.

There are no permanent surface streams, springs, or tanks. The central and eastern tiers of sections bear a good growth of grass composed of strongly rooted, tufted species of *Poa*, *Festuca*, and *Stipa*, with some patches of mesquite sward. If watering places existed the tract would be valuable for pasturage. Sheep are grazed there in the late fall and again in the early spring, but not to any considerable extent, owing to the lack of water.

Stand of timber species in T. 23 N., R. 8 E.

Feet B. M.

Yellow pine 24, 660, 000

Forest conditions in T. 23 N., R. 8 E.

Average total height ....feet.. 80
Average height, clear ....do.... 8
Average diameter, breast-high ....inches.. 18
Average age ....years.. 175
Reproduction, sections 3, 15, 16, 25, 26, 27, 28, 29, 35, and 36..................... Poor.
Reproduction, elsewhere.. Medium.

TOWNSHIP 23 NORTH, RANGE 9 EAST.

The surface features consist of extensive groups of extinct craters and cones rising 400 to 800 feet above the general level, tracts of rolling land deeply buried under vast deposits of coal-black cinders, ridges and combs of black vesicular lava rising here and there out of the mass of scoriaceous matter, and great circular or oval pits from 100 to 200 feet in depth in the rolling or level areas, from which were ejected, in part, the cinders and scoriæ which cover the region. The soil infrequently contains a small proportion of loam, but as a rule is very barren and sterile.

There is no permanent, and probably no temporary, surface flow, nor are there any springs or tanks. The township is worthless for grazing purposes, owing to its extremely sparse and scattered grass growth.

Stand of timber species in T. 23 N., R. 9 E.

	Feet B. M.
Yellow pine	6,625,000

Forest conditions in T. 23 N., R. 9 E.

Average total height	feet..	75
Average height, clear	do....	8
Average diameter, breast-high	inches..	16
Average age	years..	160
Reproduction		Poor.

TOWNSHIP 24 NORTH, RANGE 3 EAST.

The northern tiers and most of the west-central sections of this township consist of level or gently rolling areas, varied by terraces due to low breaks in the underlying volcanic rocks or to successive flows of lava. Sections 15, 23, 24, 25, 26, 31, 32, 33, 34, 35, and 36 are mostly covered with the remains of extinct craters. They occur as ranges of hills connected by ancient streams of lava, and generally lack the conoidal symmetry which distinguishes so many of the smaller eruptive foci elsewhere in the reserve. Their slopes are steep, commonly covered with red and black slag and scoriæ. A few of them are composed of tufaceous material in which erosion has scuptured deep gullies and ravines. The soil in proximity to the different cinder and slag ridges of the southern sections is composed of volcanic débris. The surface is bowlder strewn, and the top soil contains small admixtures of loamy material. In the northern sections much of the soil consists of adobe more or less mixed with coarse and fine volcanic detritus.

There are no permanent surface streams, springs, or tanks. Most of the township has been sheeped in excess of the recuperative power of the mesquite grass which here forms the bulk of the gramineous flora. The result has been an increase of coarse, valueless weeds and the extermination of the grass.

Stand of timber species in T. 24 N., R. 3 E.

	Feet B. M.
Yellow pine	95,000

Forest conditions in T. 24 N., R. 3 E.

Average total height..feet.. 75
Average height, clear ....do.... 8
Average diameter, breast-high..............................inches.. 16
Average age ....years.. 160
Reproduction..Poor.

TOWNSHIP 24 NORTH, RANGE 4 EAST.

More than one-half of this township consists of level and moderately rolling land—lava plains roughened with low hillocks and reefs of vesicular or homogeneous lava and intersected by shallow, rocky gullies and runs. Extinct volcanic cones, 350 to 650 feet high, rise in sections 2, 4, 6, 7, 8, 9, 16, 17, 18, 29, 30, 31, while sections 25, 35, and 36 chiefly comprise rough, irregular, isolated hills of lava, possibly outflows from local fissures. Most of the cones lie in groups and form short ranges of hills. The soil cover on the level and rolling areas consists of black adobe more or less mixed with fine volcanic detritus, hence forming a sort of gravelly loam. The surface is stony and bowlder strewn, often excessively so.

There is no permanent surface flow or spring. In section 34 an embankment, thrown across the opening of a gully, has raised the level of a previously existing natural pool 6 or 7 feet, resulting in a small lake, which after the spring break-up covers an area of about 100 acres and retains water throughout the year. The township practically has no present pasturage value, owing to excessive sheep herding continued through many years.

Stand of timber species in T. 24 N., R. 4 E.

Feet B. M.
Yellow pine 8,591,000

Forest conditions in T. 24 N., R. 4 E.

Average total height ....feet.. 75
Average height, clear.. do.... 8
Average diameter, breast-high ....inches.. 16
Average age ....years.. 160
Reproduction, seedling growth........................... Very poor.
Reproduction, 50 to 90 year old saplings................. Good.

TOWNSHIP 24 NORTH, RANGE 5 EAST.

The two northern tiers of sections in this township comprise rolling and hilly areas alternating with tracts of nearly level ground, and are situated at a general altitude of 7,000 feet. The central tiers of sections and the southwest quarter of the township consist of a broad, undulating terrace formed by many successive flows of lava from ancient craters of Kendrick Peak. Sections 22, 23, 25, 26, 27, and, in part, 32 and 33, form northern slopes and spurs of Kendrick Peak, which is situated in sections 34 and 35, and attains an elevation of 9,200 feet. The terrace, which constitutes most of the central portion of the township, has an elevation of 600 to 800 feet above the level and rolling areas comprised in the northern tiers of sections. It breaks off to these lower levels with a very steep and rocky front, which in sections 19, 30, and 31 abounds in precipitous escarpments and is seamed and cut by short, deep, extremely rocky canyons. Kendrick Peak is of volcanic origin, being formed by a number of ancient craters which long since lost their symmetry by erosion and now are merely heads of ravines. The soil is a gravelly loam throughout the entire township, the percentage of the gravelly constituents varying with the degree of slope—high where it is steep, low on the levels and where the declivity is gentle. The surface is commonly stony and bowlder strewn, but small tracts in sections 26 and 36 are free from surface rock.

There is no permanent surface stream in the township. In section 36 a crateriform depression on a summit of a small cinder cone holds a diminutive pool which contains water during the greater portion of the year. In the same section are two small springs which have their origin among the western spurs of Kendrick Mountain. The township has a moderate grazing value. The nontimbered areas have been sheeped until they produce nothing but coarse weeds, but the yellow pine forests have a ground cover of strongly rooted, tufted species of *Poa* and *Festuca* as yet but little grazed. The direct and higher slopes of Kendrick Peak have no grazing value, owing to the thickset stands of young timber which completely choke out all growth of grass.

Stand of timber species in T. 24, N., R. 5 E.

	Feet B. M.
Yellow pine	58,330,000
Red fir	630,000
Engelmann spruce	540,000
Arizona fir	20,000
Total	59,520,000

Forest conditions in T. 24 N., R. 5 E.

Average total height	feet..	80
Average height, clear	do....	10
Average diameter, breast-high	inches...	18
Average age	years..	180
Reproduction, yellow pine on sections 1 and 2		Good.
Reproduction, yellow pine elsewhere		Poor.
Reproduction, other species		Good.

TOWNSHIP 24 NORTH, RANGE 6 EAST.

The north tier of sections consists of rolling scab land made up of terraced lava flows which issued from craters situated in the main mass of San Francisco Peak. The two eastern tiers of sections, together with sections 28, 29, 30, 31, and

32, are mostly covered with cinder and slag cones, which in some cases are dispersed singly, but more generally occur in groups or as ranges of hills. Their height varies from 250 to 300 feet in the western areas of the township to 900 to 1,000 feet in the eastern tiers of sections. These cones show evidence of considerable erosion, here indicative of great age. Their craters have been cut down until they constitute mere heads of large gullies, and the slopes have acquired easy gradients. Between the cones lie tracts of more or less level and rolling land. The soil is gravelly loam, the gravel consisting of finely comminuted volcanic débris. In the northern tiers of sections much of the soil is black adobe with comparatively small admixture of gravel. Most of the surface is bowlder strewn or covered with great quantities of small stones derived from the underlying sheet of lava.

There are no permanent surface streams, springs, or tanks. The north tier of sections, and some of those along the east margin of the township, chiefly 12, 13, 23, and 24, have been sheeped in excess of the recuperative power of the grass and now have a low pasturage value. Elsewhere the ground supports a strong, close growth of tufted grasses, species of *Poa*, *Festuca*, and *Agropyron*, as yet but little pastured owing to lack of water. Were water obtainable in the neighborhood, the grazing value of these tracts would be high for many years to come.

Stand of timber species in T. 24 N., R. 6 E.

	Feet B. M.
Yellow pine	52,980,000

Forest conditions in T. 24 N., R. 6 E.

Average total height feet..	85
Average height, clear do....	10
Average diameter, breast-high inches..	20
Average age years..	190
Reproduction on sections 10, 11, 25, 35	Poor.
Reproduction elsewhere	Medium.

TOWNSHIP 24 NORTH, RANGE 7 EAST.

Most of the areas comprised within the limits of this township are covered with extinct volcanic cones. Between the cones lie rolling or level tracts of ground, chiefly contained in sections 1, 3, 4, 5, 15, 16, 20, 21, 22, 23, 24, and 25. The south tier of sections forms the outer margin of the northern foothill region of the San Francisco Mountains and consists of an extremely stony and rough lava terrace, rising 200 to 400 feet above the central and northern portions of the township. The front of this terrace, striking through sections 31, 32, 33, 35, and 36, is rocky and steep, and is cut by numerous stony and bowlder-strewn gullies and ravines heading in the slopes of San Francisco Peak. The cones are arranged either singly or in ranges of hills and vary in height from 350 to 900 feet, with steep slopes, and most of them have a central or sublateral crateriform depression.

The soil in the northern and most of the central portions of the township consists of volcanic detritus, more or less finely comminuted and mixed with small proportions of loam. The southern tiers of sections have a loamy soil, thinly and imperfectly covering the rough vesicular lava which underlies it, and hence is mostly composed of scab land. Much of the land in the central and northern portions of the township has the surface littered with blocks of volcanic ejecta.

. There are no permanent surface streams, springs, or tanks. The present grazing value is low, as all of the township, with the exception of the forested tracts, has been badly overgrazed, the original grass growth eaten or trampled out, and in place thereof close growths of weeds have sprung up. In the forested areas there is a moderately thick ground cover of strongly rooted, tufted species of *Poa* and *Festuca* still available for pasturage.

Stand of timber species in T. 24 N., R. 7 E.

	Feet B. M.
Yellow pine	12,985,000

Forest conditions in T. 24 N., R. 7 E.

Average total height	feet..	80
Average height, clear	do....	10
Average diameter, breast-high	inches..	18
Average age	years..	175
Reproduction on sections 20, 28, 29		Poor.
Reproduction elsewhere		Medium.

TOWNSHIP 24 NORTH, RANGE 8 EAST.

The land in this township chiefly consists of level or gently rolling areas in the eastern sections, rising in the southeastern areas to low terraces formed by lava flows from ancient craters near O'Leary Peak in the township adjoining on the south. Low cinder and slag cones rise in sections 6, 7, 8, 9, 10, 15, 16, and 19, while sections 33, 34, 35, and 36 comprise slopes and terraced lava flows flanking O'Leary Peak. The level areas are furrowed by shallow runs and gullies, while ravines and short canyons cut into the front of the lava terraces in the southeast quarter of the township. The soil is formed from comminuted volcanic débris, mixed with a small proportion of loam. Most of the southeast quarter of the township is covered with deposits of moderately fine black cinders derived from the great cinder cones around Sunset Peak in the township adjoining on the south. This cinder cover is practically without any loam admixtures on much of the land in sections 24, 25, 35, and 36, and drifts about with the wind. Elsewhere the black cinder surface is smoothed out with loamy admixtures and has acquired stability.

There are no permanent surface streams, springs, or tanks. The pasturage value is generally low, as there is no grass growth in the wooded areas, and the nontimbered tracts have, for the most part, been closely sheeped, and on the forested tracts the ground cover of tufted grasses is sparse and scattering.

Stand of timber species in T. 24 N., R. 8 E.

	Feet B. M.
Yellow pine	1,600,000

Forest conditions in T. 24 N., R. 8 E.

Average total height	feet..	75
Average height, clear	do....	8
Average diameter, breast-high	inches..	16
Average age	years..	160
Reproduction		Poor.

TOWNSHIP 24 NORTH, RANGE 9 EAST.

Sections 1 to 27, inclusive, comprise level and gently rolling tracts of land, here and there rising into low ridges and cones of lava, or terraced with thin, successive lava flows. A low cinder cone rises in section 18. Sections 28, 29, 30, and 31 consist of an extremely rough and rugged flow of lava which appears to have welled out from fissures or cones existing in the vicinity of O'Leary Peak. Sections 32 and 34 comprise low cinder cones and their slopes. The soil is composed of volcanic débris with slight loamy admixtures, or lacks these latter constituents altogether. Much of it is of the black cinder type; some of it is made up of yellow or even white scoriaceous or pumice-like material. In most localities in the township, the surface of the cinder blanket has acquired a moderate degree of stability, but in sections 19, 20, 29, and 30 the loose cindery soil drifts about with the shifting winds.

There are no permanent surface streams, springs, or tanks. The township has no value for pasturage, as the grass growth has always been extremely sparse and scattered, and sheep herding during many years has closely cleaned up the little grass originally growing there.

Stand of timber species in T. 24 N., R. 9. E.

	Feet B. M.
Yellow pine	1,380,000

Forest conditions in T. 24 N., R. 9 E.

Average total height	feet..	75
Average height, clear	do....	8
Average diameter, breast-high	inches..	16
Average age	years..	160
Reproduction		Poor.

TOWNSHIP 25 NORTH, RANGE 3 EAST.

This township consists of level or gently rolling tracts of land, in sections 24 and 25 rising into low conical hills of volcanic origin, and along the west lines of section 18, forming in part the eastern slopes of an extremely rough mesa, composed of lava, which covers most of the area of the township adjoining on the west. The southern tiers of sections are cut by shallow dry runs and intersected with reefs and combs of lava which barely rise 3 or 4 feet above the surface, while here and there low breaks in the lava sheet, which underlies the entire township, form small terraces with rocky steep fronts. The soil is composed of volcanic débris—small fragments of lava, scoriæ, and cinders—more or less mixed with blocks of loose lava, in part derived from the reefs and in part from tali of the various escarpments. Most of the surface is free from accumulations of bowlders and consists of small pebbly material with some vegetable mold.

There are no permanent surface streams, springs, or tanks. There is a woodland growth of piñon and one-seed juniper, of which the reproduction is poor. The grazing value of the township is low, as owing to aridity the grass is short and scattering even at its best. It is chiefly mesquite grass, *Bouteloua oligostacha.* The southern sections have been pastured fairly close by sheep.

There is no timber.

TOWNSHIP 25 NORTH, RANGE 4 EAST.

Most of the land in this township is level or gently rolling. Sections 10, 11 (in part), 12, 13, 16, 17, and 18 contain low extinct volcanic cones, situated singly or in groups, cindery or tufaceous in character, none covering a large area. The slopes of these cones are usually steep. A few contain a central depression at the top. Mostly, however, the craters are lateral. In sections 14 and 15 the surface features are rather more rough and rolling than elsewhere. The soil is generally composed of volcanic débris—scoriæ and cinders—the surface often strewn with blocks of vesicular lava and slaggy masses of the same. It is a barren soil containing only a very small proportion of vegetable mold, while humus is wholly lacking.

There are no permanent surface streams, springs, or tanks. The wooded areas bear stands of piñon and one-seed juniper, with some underbrush consisting of mountain mahogany, rock rose, and other less conspicuous herbs of the semiarid region of northern Arizona. Reproduction is slow. Most of the township has been closely sheeped and its value for grazing is low.

There is no timber.

TOWNSHIP 25 NORTH, RANGE 5 EAST.

The northern, western, and portions of the southern sections of this township consist of level or gently rolling areas, varied by low breaks in the lava sheet or intersected by shallow, rocky ravines. Sections 20, 21, 28, 29, 34, and 35 are covered wholly or in part with the remains of extinct volcanic cones, usually low, but occasionally, as in sections 28 and 29, rising 1,000 feet above the plateau level. In most places throughout the township the soil is composed of more or less finely comminuted volcanic débris, some of which consists of coarsely pulverized slag or scoriæ.

There are no permanent streams. A spring exists in the crater of a cinder cone situated in sections 20 and 21, but it is of small capacity, a mere trickle during the dry portion of the year and probably fails altogether some years. The northern and central portions of the township have been only moderately sheeped, but the southern sections are badly overgrazed and in places the grass has been completely destroyed.

Stand of timber species in T. 25 N., R. 5 E.

Feet B. M.

Yellow pine 2,572,000

Forest conditions in T. 25 N., R. 5 E.

Average total height ....feet.. 75
Average height, clear ....do.... 8
Average diameter, breast-high ....inches.. 16
Average age ....years.. 160
Reproductions, sections 20, 21, 28, 29, 34, 35, 36 .... Medium.
Reproduction elsewhere .... Poor.

TOWNSHIP 25 NORTH, RANGE 6 EAST.

The eastern sections comprise land nearly level or gently rolling. The northern tiers of sections consist of a low mesa formed of limestone, thinly capped with lava in some places, deeply buried under outflows of volcanic matter in others. It is cut by shallow, rocky runs and gullies, which occasionally develop a canyon formation where they open into the level areas of the eastern sections. The southern portion of the township consists of a terrace or mesa of vesicular and scoriaceous lava. Along a line running through sections 18, 17, 16, 21, 26, and 35 this terrace breaks off with a steep, rocky front, 250 to 400 feet high, to the lower areas of the northern sections of the township. The terrace is very rocky and bowlder strewn, and is cut at frequent intervals by gullies and short canyons. Cinder and slag cones, marking ancient volcanic vents, rise in the southwest quarter of section 16, and in sections 30 and 31. The cones are low, none above 250 feet high, with gentle slopes which are strewn with red and black scoriæ, slag, and cinders ejected from the ancient craters of these cones. The northern portion of the township has a sandy or gravelly soil largely derived from the decomposition of the limestone outcrops in the two northern tiers of sections. The soil in the southern areas consists of adobe, or where mixed with volcanic cinders it assumes the character of gravelly loam. The surface of the eastern sections is, as a rule, smooth and pebbly.

There are no permanent surface streams. There are four small springs in section 20 and one in section 28 which rise through crevices in the lava along the foot of the escarpment. Their flow is insignificant, and is used for watering stock.

The woodland contains piñon and one-seed juniper, with a few yellow pine, red fir, walnut, cottonwood, and willows. The grazing value is low.

There is no timber.

TOWNSHIP 25 NORTH, RANGE 7 EAST.

The most conspicuous features of the township are the numerous extinct volcanic cones. One or more rises in every section except sections 26 and 27. Some of these cones stand as isolated buttes, others are grouped or occur as ranges of hills varying in height from mere slag heaps 10 to 50 feet high, to rounded symmetrical truncated cones 400 to 800 feet high. Generally they possess a distinct circular depression on their summits. Sometimes the depression is lateral, while in other cases the rims of the ancient craters have been largely removed by erosion and all that remains are the heads of deep gullies cutting into the cones. The cones and hills are separated by level tracts mostly of small extent. The soil consists almost wholly of volcanic débris. On the slopes of the cones the soil is very thin, often entirely lacking, in which case the surface is covered with coarse fragments of slag, scoriæ, and vesicular lava. On the level areas separating the hills the volcanic débris is finer and the soil of a gravelly character, occasionally topped with an accumulation of loam.

There are no permanent surface streams, springs, or tanks. The woodland contains piñon, one-seed juniper, and a few scattered yellow pines. Reproduction is slow and uncertain. The land has been closely grazed by sheep and cattle. Much of the original growth of mesquite grass has been destroyed, and coarse, valueless weeds have come in its place. The pasturage value is very low.

There is no timber.

TOWNSHIP 25 NORTH, RANGE 8 EAST.

Most of this township consists of level or gently rolling land. In sections 17, 18, 19, 20, 29, 30, 31, 34, and 35 rise low extinct volcanic cones, some of which occur singly, others in groups. They vary in height from 400 to 900 feet, with steep slopes, and generally with well-defined crateriform depressions on their summits. The soil throughout is composed of volcanic ejecta. On the slopes and in the vicinity of the cones it is slaggy or clinker like. On the level areas and at some distance from the ancient vents the volcanic débris is more comminuted and the soil is distinctly of a gravelly or pebbly character.

There are no permanent surface streams, springs, or tanks. There is a woodland growth of piñon and one-seed juniper in which the reproduction is limited. The pasturage value is very low. It has been grazed so long and close by sheep, cattle, and horses that the growth of mesquite grass, never very luxuriant, is practically killed, and coarse and valueless weeds have come in and occupied the ground. Most of the sheeping and stock grazing in this township and adjoining ones takes place during the winter months when the melting of the snow and the occasional rains supply the necessary water.

There is no timber.

TOWNSHIP 25 NORTH, RANGE 9 EAST.

Most of the land in the township is gently rolling. The entire tract is underlain with lava flows of various ages. Successive flows, stopping short of the points reached by preceding ones, have caused the formation of irregular terraces usually presenting low rocky or cinder-covered fronts. Shallow runs intersect the township in various directions, and in sections 13, 14, 15, 16, 20, 23, and 24 rise low extinct volcanic cones which are commonly asymmetrical in outline, owing to erosion and to successive eruptions through different orifices of the same cone. The soil is composed of more or less finely comminuted volcanic débris mixed with small proportions of loamy constituents. A great deal of the soil consists of lapilli ejected from the great cinder cones in T. 23 N., R. 8 E., about 12 miles to the southwest, having been transported either by water or wind. The general quality of the soil might be characterized as a gravelly or sandy loam, but it is decidedly barren and sterile.

There are no permanent surface streams, springs, or tanks. The woodland contains piñon and one-seed juniper, of which the reproduction is slow and uncertain. The pasturage value of the tract is low.

There is no timber.

SUMMARY.

Classification of lands in San Francisco Mountains Forest Reserve, by townships.

Township.	Range.	Timbered area.	Wooded area.	Burned area.	Cut area.a	Barren area.b
		Acres.	*Acres.*	*Acres.*	*Acres.*	*Acres.*
15 north.............	6 east.................	....	7,040	....		4,480
15 north.............	7 east.................	....	15,360	....		
15 north.............	8 east.................	6,400	8,960	....		...
15 north.............	9 east.................	11,520	....	....		
15 north.............	10 east................	2,600	1,240	....		
15 north.............	11 east................	....	3,840	....		
16 north.............	5 east.................	....	3,640	....		19,400
16 north.............	6 east.................	....	9,600	....		13,440
16 north.............	7 east.................	....	10,880.	....		12,160
16 north.............	8 east.................	15,400	7,100	....		540
16 north.............	9 east.................	23,040	....	....		

a Logged more than 65 per cent or until remaining stand is too small to be classed as timber.
b Naturally timberless areas, including lakes.

Classification of lands in San Francisco Mountains Forest Reserve, by townships—Continued.

Township.	Range.	Timbered area.	Wooded area.	Burned area.	Cut area.	Barren area.
		Acres.	*Acres.*	*Acres.*	*Acres.*	*Acres.*
16 north	10 east	8,300	14,740			
16 north	11 east		22,500			540
17 north	4 east					23,040
17 north	5 east					23,040
17 north	6 east		23,040			
17 north	7 east	8,400	9,100			
17 north	8 east	21,100	1,940			
17 north	9 east	23,040				
17 north	10 east	12,800	10,240			
17 north	11 east		23,040			
18 north	4 east					23,040
18 north	5 east	6,000	4,000			13,040
18 north	6 east	7,680	15,360			
18 north	7 east	16,100	1,400			
18 north	8 east	23,040				
18 north	9 east	8,300	6,700			8,040
18 north	10 east	3,000	20,040			
18 north	11 east		23,040			
19 north	1 east		23,040			
19 north	2 east	6,000	17,040			
19 north	3 east	21,100	1,940			
19 north	4 east	23,040				
19 north	5 east	23,040				
19 north	6 east	25,680				
19 north	7 east	20,400				
19 north	8 east	20,480			2,560	
19 north	9 east	15,300	7,000			740
19 north	10 east		23,040			
20 north	1 east	6,500	16,540			
20 north	2 east	19,500	2,900		640	
20 north	3 east	17,500			5,540	
20 north	4 east	23,040				
20 north	5 east	23,040				
20 north	6 east	21,570			1,470	
20 north	7 east	8,440			14,420	180
20 north	8 east	7,700	6,400		8,200	740
20 north	9 east	1,000	22,040			
20 north	10 east		23,040			
21 north	1 east	5,900	17,140			
21 north	2 east	18,080			4,960	

Classification of lands in San Francisco Mountains Forest Reserve, by townships—Continued.

Township.	Range.	Timbered area.	Wooded area.	Burned area.	Cut area.	Barren area.
		Acres.	*Acres.*	*Acres.*	*Acres.*	*Acres.*
21 north............	3 east.................	9,300	....	....	12,975	765
21 north............	4 east..................	7,885	....	....	1,440	13,715
21 north............	5 east.................	18,770	....	....	4,270	....
21 north............	6 east.................	15,215	....	....	7,825	....
21 north............	7 east.................	11,215	585	....	11,240	....
21 north............	8 east.................	14,825	5,715	....	2,500	....
21 north............	9 east.................	....	17,920	4,500	....	620
22 north............	1 east.................	1,500	2,500	....	....	19,040
22 north............	2 east.................	3,800	19,240	....	....	....
22 north............	3 east.................	4,390	135	....	2,570	15,945
22 north............	4 east.................	8,305	....	....	8,870	5,865
22 north............	5 east.................	17,480	....	....	5,170	390
22 north............	6 east.................	18,900	....	....	2,720	1,420
22 north............	7 east.................	21,040	315	200	1,485	....
22 north............	8 east.................	13,960	1,545	....	1,000	6,535
22 north............	9 east.................	4,585	18,075	....	....	380
23 north............	3 east.................	6,405	2,590	40	....	14,005
23 north............	4 east.................	15,930	290	....	50	6,770
23 north............	5 east.................	16,220	....	50	....	6,770
23 north............	6 east.................	16,970	200	250	....	5,620
23 north............	7 east.................	17,870	750	1,200	....	3,220
23 north............	8 east.................	11,610	4,190	....	....	7,240
23 north............	9 east.................	3,790	14,730	....	....	4,520
24 north............	3 east.................	740	21,610	....	....	690
24 north............	4 east.................	5,845	6,360	....	....	10,835
24 north............	5 east.................	16,285	2,250	80	....	4,425
24 north............	6 east.................	14,350	4,480	....	....	4,210
24 north............	7 east.................	5,085	8,235	....	....	9,720
24 north............	8 east.................	2,470	9,890	100	....	10,580
24 north............	9 east.................	2,240	20,800	....	....	....
25 north............	3 east.................	....	23,040	....	....	....
25 north............	4 east.................	....	15,960	....	....	7,080
25 north............	5 east.................	1,490	19,805	120	....	1,625
25 north............	6 east.................	....	15,360	250	....	7,430
25 north............	7 east.................	....	8,320	....	....	14,720
25 north............	8 east.................	....	1,280	....	....	21,760
25 north............	9 east.................	....	9,280	....	....	13,760
		812,500	658,370	6,790	99,905	362,075

Total area of reserve 1,939,640 acres.

Stand and classification of merchantable timber in San Francisco Mountains Forest Reserve, by townships.

Township.	Range.	Total stand.	Average stand per acre.	Yellow pine.	Red fir.	White fir.	Engelmann spruce.	Other species.
		Feet B. M.	*Feet B. M.*	*Feet B. M.*	*Feet B. M.*	*Feet B. M*	*Feet B. M.*	*Feet B. M.*
15 north..	6 east...							
15 north..	7 east...							
15 north..	8 east...	6,500,000	1,016	6,500,000				
15 north..	9 east...	17,000,000	1,467	17,000,000				
15 north..	10 east..	4,000,000	1,539	4,000,000				
15 north..	11 east..							
16 north..	5 east...							
16 north..	6 east...							
16 north..	7 east...							
16 north..	8 east...	34,000,000	2,210	34,000,000				
16 north..	9 east...	66,500,000	2,886	66,500,000				
16 north..	10 east..	12,250,000	1,475	12,250,000				
16 north..	11 east..							
17 north..	4 east...							
17 north..	5 east...							
17 north..	6 east...							
17 north..	7 east...	15,500,000	1,845	15,500,000				
17 north..	8 east...	71,000,000	3,361	71,000,000				
17 north..	9 east...	80,000,000	3,472	80,000,000				
17 north..	10 east..	38,250,000	3,000	38,250,000				
17 north..	11 east..							
18 north..	4 east...							
18 north..	5 east...	11,750,000	1,958	11,750,000				
18 north..	6 east...	13,500,000	1,758	13,500,000				
18 north..	7 east...	55,250,000	3,432	55,250,000				
18 north..	8 east...	106,750,000	4,633	105,000,000	1,750,000			
18 north..	9 east...	24,250,000	3,000	24,250,000				
18 north..	10 east..	4,000,000	1,334	4,000,000				
18 north..	11 east..							
19 north..	1 east...							
19 north..	2 east...	13,500,000	2,250	13,500,000				
19 north..	3 east...	81,500,000	3,768	81,500,000				
19 north..	4 east...	54,250,000	2,355	54,250,000				
19 north..	5 east...	85,750,000	3,722	85,750,000				
19 north..	6 east...	78,750,000	3,418	78,750,000				
19 north..	7 east...	91,250,000	3,965	91,250,000				
19 north..	8 east...	56,000,000	2,734	56,000,000				
19 north..	9 east...	23,500,000	1,536	23,500,000				
19 north..	10 east..							
20 north..	1 east...	9,250,000	1,423	9,250,000				
20 north..	2 east...	69,000,000	3,538	69,000,000				

Stand and classification of merchantable timber in San Francisco Mountains Forest Reserve, by townships—
Continued.

Township.	Range.	Total stand.	Average stand per acre.	Yellow pine.	Red fir.	White fir.	Engelmann spruce.	Other species.
		Feet B. M.	*Feet B. M.*	*Feet B. M.*	*Feet B. M.*	*Feet B. M.*	*Feet B. M.*	*Feet B. M.*
20 north..	3 east...	82,000,000	4,685	82,000,000				
20 north..	4 east...	97,500,000	4,210	97,500,000				
20 north..	5 east...	151,500,000	6,575	151,500,000				
20 north..	6 east...	94,500,000	4,381	94,500,000				
20 north..	7 east...	50,340,000	5,340	50,340,000				
20 north..	8 east...	24,000,000	3,118	24,000,000				
20 north..	9 east...	1,000,000	1,000	1,000,000				
20 north..	10 east..							
21 north..	1 east...	8,250,000	1,400	8,250,000				
21 north..	2 east...	36,000,000	1,562	36,000,000				
21 north..	3 east...	40,335,000	4,337	40,335,000				
21 north..	4 east...	39,350,000	4,990	39,350,000				
21 north..	5 east...	109,520,000	5,835	109,520,000				
21 north..	6 east...	42,275,000	2,779	42,275,000				
21 north..	7 east...	33,040,000	2,946	33,040,000				
21 north..	8 east...	42,778,000	2,886	42,778,000				
21 north..	9 east...							
22 north..	1 east...	750,000	500	750,000				
22 north..	2 east...	5,000,000	1,315	5,000,000				
22 north..	3 east...	15,074,000	3,434	15,074,000				
22 north..	4 east...	19,849,000	2,390	19,849,000				
22 north..	5 east...	79,320,000	4,538	79,320,000				
22 north..	6 east...	82,416,000	4,361	82,416,000				
22 north..	7 east...	42,480,000	2,002	36,108,000	2,124,000	3,398,000	850,000	
22 north..	8 east...	51,435,000	3,684	51,178,000	128,000	129,000		
22 north..	9 east...	9,730,000	2,122	9,730,000				
23 north..	3 east...	24,135,000	3,768	24,135,000				
23 north..	4 east...	68,408,000	4,445	65,672,000	2,394,000	171,000		171,000
23 north..	5 east...	76,410,000	4,711	74,729,000	100,000	40,000	1,528,000	13,000
23 north..	6 east...	78,185,000	4,607	77,794,000	391,000			
23 north..	7 east...	43,720,000	2,447	39,827,000	919,000	1,443,000	1,531,000	
23 north..	8 east...	24,660,000	2,124	24,660,000				
23 north..	9 east...	6,625,000	1,748	6,625,000				
24 north..	3 east...	95,000	28	95,000				
24 north..	4 east...	8,591,000	1,470	8,591,000				
24 north..	5 east...	59,520,000	3,655	58,330,000	630,000		540,000	20,000
24 north..	6 east...	52,980,000	3,700	52,980,000				
24 north..	7 east...	12,985,000	2,554	12,985,000				
24 north..	8 east...	1,600,000	648	1,600,000				
24 north..	9 east...	1,380,000	616	1,380,000				
25 north..	3 east...							

Stand and classification of merchantable timber in San Francisco Mountains Forest Reserve, by townships—
Continued.

Township.	Range.	Total stand.	Average stand per acre.	Yellow pine.	Red fir.	White fir.	Engelmann spruce.	Other species.
		Feet B. M.	*Feet B. M.*	*Feet B. M.*	*Feet B. M.*	*Feet B. M.*	*Feet B. M.*	*Feet B. M.*
25 north..	4 east...							
25 north..	5 east...	2,572,000	1,726	2,572,000				
25 north..	6 east...							
25 north..	7 east...							
25 north..	8 east...							
25 north..	9 east...							
Total.		2,743,558,000		2,725,288,000	8,436,000	5,181,000	4,449,000	204,000

Percentages of merchantable and nonmerchantable species in San Francisco Mountains Forest Reserve, by townships.

Township.	Range.	Yellow pine.	Limber pine.	Piñon.	Red fir.	Arizona fir.	White fir.	Engelmann spruce.	Alligator juniper.	One-seed juniper.	Aspen.	Oak.	Arizona cypress.	Other species.
15 north....	6 east													
15 north....	7 east												100	....
15 north....	8 east	60										6	34	....
15 north....	9 east	97										3		
15 north....	10 east	80										15	5	....
15 north....	11 east												100	....
16 north....	5 east													
16 north....	6 east													
16 north....	7 east													
16 north....	8 east	90										3	7	
16 north....	9 east	86										14		
16 north....	10 east	37										13	50	....
16 north....	11 east												100	....
17 north....	4 east													
17 north....	5 east													
17 north....	6 east											25	75	....
17 north....	7 east	71										14	15	....
17 north....	8 east	84										1	15	....
17 north....	9 east	91										9		
17 north....	10 east	85										11	4	....
17 north....	11 east												100	....
18 north....	4 east													
18 north....	5 east	89											11	....

Percentage of merchantable and nonmerchantable species in San Francisco Mountains Forest Reserve, by townships—Continued.

Township.	Range.	Yellow pine.	Limber pine.	Piñon.	Red fir.	Arizona fir.	White fir.	Engelmann spruce.	Alligator juniper.	One-seed juniper.	Aspen.	Oak.	Arizona cypress.	Other species.
18 north	6 east	58										27	15	
18 north	7 east	83										15	2	
18 north	8 east	90			2							8		
18 north	9 east	84										12	4	
18 north	10 east	50											50	
18 north	11 east												100	
19 north	1 east												100	
19 north	2 east	41										1	58	
19 north	3 east	88										8	4	
19 north	4 east	84										15	1	
19 north	5 east	92										5	3	
19 north	6 east	79										21		
19 north	7 east	94										6		
19 north	8 east	97										3		
19 north	9 east	60										9	31	
19 north	10 east											1	99	
20 north	1 east	73										7	20	
20 north	2 east	93										6	1	
20 north	3 east	92										8		
20 north	4 east	95										5		
20 north	5 east	96										4		
20 north	6 east	96										x		
20 north	7 east	100												
20 north	8 east	77										5	18	
20 north	9 east	1										2	97	
20 north	10 east												100	
21 north	1 east	73										2	25	
21 north	2 east	97										3		
21 north	3 east	98										2		
21 north	4 east	100												
21 north	5 east	96										4		
21 north	6 east	98										2		
21 north	7 east	95			1		2					2		
21 north	8 east	91		6						1	1	1		
21 north	9 east													
22 north	1 east	95											5	
22 north	2 east	90										9	1	
22 north	3 east	100												

Percentages of merchantable and nonmerchantable species in San Francisco Mountains Forest Reserve, by townships—Continued.

Township.	Range.	Yellow pine.	Limber pine.	Piñon.	Red fir.	Arizona fir.	White fir.	Engelmann spruce.	Alligator juniper.	One-seed juniper.	Aspen.	Oak.	Arizona cypress.	Other species.
22 north	4 east	100												
22 north	5 east	100												
22 north	6 east	72			1						27			
22 north	7 east	61	2		7	2	13	6			9			
22 north	8 east	97		1			1							
22 north	9 east	77		18						5				
23 north	3 east	95		2	2					1				
23 north	4 east	88	3		3		1				5			
23 north	5 east	80	1		2		2	7			8			
23 north	6 east	53	2		7	1		5			32			
23 north	7 east	69	1		2	1	3	10			14			
23 north	8 east	70		22						8				
23 north	9 east	65		15						20				
24 north	3 east	30		48						22				
24 north	4 east	94		5						1				
24 north	5 east	83		1	5			2	1		8			1
24 north	6 east	98		1										
24 north	7 east	84		14						2				
24 north	8 east	90		7						3				
24 north	9 east	90		2						8				
25 north	3 east													
25 north	4 east													
25 north	5 east	65		20					3	12				
25 north	6 east													
25 north	7 east													
25 north	8 east													
25 north	9 east													

Composition of the nonmerchantable arborescent growth on the wooded areas of San Francisco Mountains Forest Reserve, by townships and percentages.

Township.	Range.	Yellow pine.	Limber pine.	Piñon.	Alligator juniper.	One-seed juniper.	Aspen.	Oak.	Arizona cypress.	Other species.
15 north....	6 east					99				1
15 north....	7 east								100	
15 north....	8 east	x						x		
15 north....	9 east	x						x		
15 north....	10 east	x						x	x	
15 north....	11 east								x	
16 north....	5 east					99				x
16 north....	6 east					99				1
16 north....	7 east					99	x			1
16 north....	8 east	x						x		
16 north....	9 east	x					x	x	x	
16 north....	10 east								x	
16 north....	11 east				x				x	
17 north....	4 east									
17 north....	5 east				x			x	x	x
17 north....	6 east							x		x
17 north....	7 east							x	x	
17 north....	8 east	x						x		
17 north....	9 east	x						x	x	
17 north....	10 east	x							x	
17 north....	11 east								x	
18 north....	4 east									
18 north....	5 east							x	x	
18 north....	6 east	x						x	x	x
18 north....	7 east	x						x	x	
18 north....	8 east	x						x		
18 north....	9 east	x						x	x	
18 north....	10 east	x							x	
18 north....	11 east				x				x	
19 north....	1 east								x	
19 north....	2 east	x						x	x	
19 north....	3 east	x						x		
19 north....	4 east	x						x		
19 north....	5 east	x						x		
19 north....	6 east	x						x		x
19 north....	7 east	x						x		
19 north....	8 east	x						x		
19 north....	9 east	x						x		
19 north....	10 east	x							x	
20 north....	1 east	x					x	x		

(x) Represented.

Composition of the nonmerchantable arborescent growth on the wooded areas of San Francisco Mountains Forest Reserve, by townships and percentages—Continued.

Township.	Range.	Yellow pine.	Limber pine.	Piñon.	Alligator juniper.	One-seed juniper.	Aspen.	Oak.	Arizona cypress.	Other species.
20 north	2 east	x					x	x	x	
20 north	3 east	x						x		
20 north	4 east	x					x	x		
20 north	5 east	x						x		
20 north	6 east									
20 north	7 east									
20 north	8 east							x	x	
20 north	9 east							x	x	
20 north	10 east								x	
21 north	1 east							x	x	
21 north	2 east	x					x	x		
21 north	3 east									
21 north	4 east									
21 north	5 east									
21 north	6 east									
21 north	7 east			90	x	10				
21 north	8 east	1		44		55				x
21 north	9 east			40		60				
22 north	1 east	x						x		
22 north	2 east							x		
22 north	3 east			90		10				
22 north	4 east			ı	x					
22 north	5 east									
22 north	6 east									
22 north	7 east		1				99			x
22 north	8 east	7		85		8				
22 north	9 east	2		48		50				
23 north	3 east			75		25				
23 north	4 east	1		97		2				
23 north	5 east									
23 north	6 east		x				100			
23 north	7 east		x				100			
23 north	8 east		x	65		35				
23 north	9 east			40		60				
24 north	3 east	1		74		25				
24 north	4 east			71		29				
24 north	5 east	x		85		15				
24 north	6 east			90		10				
24 north	7 east			65		35				
24 north	8 east			25		75				

(x) Represented.

Composition of the nonmerchantable arborescent growth on the wooded areas of San Francisco Mountains Forest Reserve, by townships and percentages—Continued.

Township.	Range.	Yellow pine.	Limber pine.	Piñon.	Alligator juniper.	One-seed juniper.	Aspen.	Oak.	Arizona cypress.	Other species.
24 north....	9 east	....	....	30	....	70	....	....	......	......
25 north....	3 east	....	....	62	....	38	....	....	......	......
25 north....	4 east	....	....	80	...,....	20	....	....	......	......
25 north....	5 east	....	....	85	x	15	....	....	......	....
25 north....	6 east	1	....	84	....	15	....	....	......	......
25 north....	7 east	x	....	70	....	30	....	....	......	......
25 north....	8 east	....	....	70	....	30	....	....	......	......
25 north....	9 east	....	....	30	....	70	....	....	......	......

(x) Represented.

INDEX.

O

PUBLICATIONS OF UNITED STATES GEOLOGICAL SURVEY

[Professional Paper No. 22.]

The serial publications of the United States Geological Survey consist of (1) Annual Reports, (2) Monographs, (3) Professional Papers, (4) Bulletins, (5) Mineral Resources, (6) Water-Supply and Irrigation Papers, (7) Topographical Atlas of the United States—folios and separate sheets thereof, (8) Geologic Atlas of the United States—folios thereof. The classes numbered 2, 7, and 8 are sold at cost of publication; the others are distributed free. A circular giving complete lists may be had on application.

The Professional Papers, Bulletins, and Water-Supply Papers treat of a variety of subjects, and the total number issued is large. They have therefore been classified into the following series: A, Economic geology; B, Descriptive geology; C, Systematic geology and paleontology; D, Petrography and mineralogy; E, Chemistry and physics; F, Geography; G, Miscellaneous; H, Forestry; I, Irrigation; J, Water storage; K, Pumping water; L, Quality of water; M, General hydrographic investigations; N, Water power; O, Underground waters; P, Hydrographic progress reports. This paper is the seventh in series H, the complete list of which follows (all are Professional Papers thus far):

SERIES H, FORESTRY.

4. The forests of Oregon, by Henry Gannett. 1902. 36 pp., 7 pls.

5. The forests of Washington, a revision of estimates, by Henry Gannett. 1902. 38 pp., 1 pl.

6. Forest conditions in the Cascade Range, Washington, between the Washington and Mount Rainier forest reserves, by F. G. Plummer. 1902. 42 pp., 11 pls.

7. Forest conditions in the Olympic Forest Reserve, Washington, from notes by Arthur Dodwell and T. F. Rixon. 1902. 110 pp., 20 pls.

8. Forest conditions in the northern Sierra Nevada, California, by J. B. Leiberg. 1902. 194 pp., 12 pls.

9. Forest conditions in the Cascade Range Forest Reserve, Oregon, by A. D. Langille, F. G. Plummer, Arthur Dodwell, T. F. Rixon, and J. B. Leiberg, with an introduction by Henry Gannett. 1903. 298 pp., 41 pls.

22. Forest conditions in the San Francisco Mountains Forest Reserve, Arizona, by J. B. Leiberg, T. F. Rixon, and Arthur Dodwell, with an introduction by F. G. Plummer. 1904. 95 pp., 7 pls.

Besides the foregoing, three volumes on forestry have been published, as Pt. V of the Nineteenth, Twentieth, and Twenty-first annual reports, each consisting of several papers.

Correspondence should be addressed to—

THE DIRECTOR,
UNITED STATES GEOLOGICAL SURVEY,
WASHINGTON, D. C.

MAY, 1904.

[Mount each slip upon a separate card, placing the subject at the top of the second slip. The name of the series should not be repeated on the series card, but additional numbers should be added, as received, to the first entry.]

Leiberg, John B.

. . . Forest conditions in the San Francisco Mountains forest reserve, Arizona, by John B. Leiberg, Theodore F. Rixon, and Arthur Dodwell, with an introduction by F. G. Plummer. Washington, Gov't print. off., 1904.

95, III p. 7 pl. (incl. map in pocket, diagr.) 29½ x 23ᶜᵐ. (U. S. Geological survey. Professional paper no. 22.)
Subject series: H, Forestry, 7.

Leiberg, John B.

. . . Forest conditions in the San Francisco Mountains forest reserve, Arizona, by John B. Leiberg, Theodore F. Rixon, and Arthur Dodwell, with an introduction by F. G. Plummer. Washington, Gov't print. off., 1904.

95, III p. 7 pl. (incl. map in pocket, diagr.) 29½ x 23ᶜᵐ. (U. S. Geological survey. Professional paper no. 22.)
Subject series: H, Forestry, 7.

U. S. Geological survey.

Professional papers.

no. 22. Leiberg, J. B. Forest conditions in the San Francisco Mountains forest reserve, Ariz., by J. B. Leiberg, T. F. Rixon, and A. Dodwell, with an introduction by F. G. Plummer. 1904

U. S. Dept. of the Interior.

see also

U. S. Geological survey.

CPSIA information can be obtained
at www.ICGtesting.com
Printed in the USA
BVHW04*1140200818
525056BV00010B/479/P